SHIPS, MACHINERY, AND MOSSBACKS

SHIPS MACHINERY AND MOSSBACKS

THE AUTOBIOGRAPHY OF A NAVAL ENGINEER

BY HAROLD G. BOWEN

VICE ADMIRAL, USN (RET.)

PRINCETON, NEW JERSEY

PRINCETON UNIVERSITY PRESS

1954

London: Geoffrey Cumberlege, Oxford University Press
L.C. CARD 53-6380

This book has been cleared by the Security Review Branch, Department of Defense, but it does not necessarily represent the official views of the Navy Department.

Printed in the United States of America
by Princeton University Press, Princeton, New Jersey

Foreword

BY FLEET ADMIRAL WILLIAM D. LEAHY,
UNITED STATES NAVY

Vice Admiral Harold G. Bowen, after forty-six years of active service in the Navy of the United States, has written this fascinating and instructive engineering autobiography in which he relates his personal difficulties and his accomplishments in the astonishing development of marine engineering since he first went to sea in 1905.

His difficulties with the very primitive propulsive installations of our ships in the early days was experienced by all of us who were his contemporaries, but being gifted with a natural engineering talent and with persistent energy his approach to these difficulties was very different from that of most of us.

While we others who survived just managed somehow to get the disabled craft back into a safe harbor and turn it over, with much relief, to such repair facilities as could be found, he directed his personal energies toward correcting the faulty engineering with the purpose of making future failures less likely.

Since his first commissioning I have watched with envy his constant devotion to and his care and improvement of the engineering equipment of the Navy; and together with all the contemporary officer personnel have admired his fearless persistence and have shared with the sea defenses the results of his continuing process of successful accomplishment.

Vice Admiral Bowen from the beginning of his naval career possessed those qualities essential to leadership; devotion to the profession of engineering; persistence in its pursuit; and that rare quality of leadership—selfless loyalty to those with whom he worked with their resulting complete and devoted loyalty to him.

The above is a Navy definition of qualities of leadership

which I have no doubt are equally valuable in the successful accomplishment of any other difficult undertaking.

We have had in the Navy during my time many successful engineering officers, but insofar as I know, no other than Vice Admiral Bowen made so many improvements, large and small in the engineering installations of ships of war, or obtained such a continuing series of rewards in the confidence and esteem of his contemporaries and the appreciation of the high command, both civil and naval, under which he worked.

His progress through the naval ranks was, because of his superior accomplishments, as rapid as is permitted by legal restrictions applicable to naval engineering officer personnel.

He served as an Assistant Engineer at sea and on shore;

as Chief Engineer of ships at sea;

as Chief Engineer of a Navy Yard;

as Fleet Engineer of the Battle Fleet;

as Chief of the Bureau of Engineering in the Navy Department, with the rank of Rear Admiral;

in charge of the Naval Research Laboratory;

in charge of the Office of Research and Inventions and the first Chief of the Office of Naval Research in the Navy Department with the rank of Vice Admiral;

awarded the Distinguished Service Medal for original and highly successful accomplishments in the Navy.

Upon reaching the statutory retirement age on 1 June 1947, after 46 years of service, he was retired as a Vice Admiral of the United States Navy.

To those who were involved to any extent in the miraculous development of engineering in the past half century this engineering autobiography is a fascinating story written by a successful engineer who is gifted with a scientific mind.

To those of a new generation who will replace us in the future development at sea and on shore of scientific engineering that is already dimly discerned, this story of Admiral Bowen's difficulties and successes should be used as a text book.

Preface

THE author of this book happened to live in and be active in the first half of this century. This period marks the rise of technology and the emergence of our country as the leader in industrialism and of Western Civilization. While Naval development borrowed much from technology and industry, it also contributed much in return. The Navy led in steam and many other phases of engineering, pioneered in electronics and, for a long time, was the only and always the staunchest supporter in the Government of basic and applied research. The Navy also pioneered in plant management (ship building) and plant seizure and operation in time of war.

All these matters should be recorded for the benefit of future students of this period.

H.G.B.

Contents

SHIPS, MACHINERY, AND MOSSBACKS

1

Education, Ashore and Afloat

The U.S. Naval Academy in 1901, to which I had won an appointment by competitive examination, was quite a letdown in many ways after my experiences at the Classical High School in Providence, Rhode Island. The Academy was in the process of being rebuilt to house a large increase in the battalion of Midshipmen. In fact, my class constituted about half of the battalion. We were housed in a temporary building, frame construction, which filled our needs and no more.

Sworn in at noon, I was standing on my head at night on board the battleship *Indiana*, and I regretted that my name began with a "B" because we were hazed in alphabetical order.

The first half of the summer cruise was spent on the *Indiana*. Hazing was good-natured. I was known as the plebe with the broad "A" and frequently was required to read books out loud to upper classmen to help them develop the accent and pronunciation of those who are born and reared northeast of the Hudson.

The last half of the cruise was spent on the *Chesapeake*, a training ship for Midshipmen. She was a three-masted full-rigged ship, and I have always remembered my experiences on that vessel with appreciation.

We had hardly gotten aboard the *Chesapeake* when lists were posted showing our stations at sailing. When the Bosun's Mate piped us to our stations for getting under way, I was dismayed not to find my name on the station bill for all hands. What should I do? If I stayed and waited to find out where my station was, everyone would think I was afraid to go aloft, so I bolted with a number of Midshipmen for an unknown destination. Better to be upbraided for doing the wrong thing than to be suspected of being a coward. The destination turned out to be, of all places for one who couldn't swim a stroke, the flying jib boom; but, as I learned later, a deep-sea sailor wasn't supposed to swim—he was supposed to drown. Furling the flying jib in the blackness of night in a furious thunder squall, the ship pitching so violently that your feet and the footrope almost touched the white-caps underneath, the jib flapping so violently it was almost impossible to hold, was a good experience for an adolescent male. On sailing ships, the tradition was one hand for yourself and one hand for the government. That is all right for work on the yards, but it won't work with that violent, temperamental, dominating flying jib. You sink your teeth into the sail as close to the ridge rope as you can and you furl that sail with both hands just as fast as you can before it sweeps you overboard.

On my last cruise on the *Chesapeake* I had graduated to captain of the mizzen t'gallant and thoroughly enjoyed myself when we were sending up and down light yards.

I think it was a great mistake to abolish the Midshipmen's training on square riggers. One cruise was on Admiral Farragut's old flagship, the *Hartford*. I can never forget her main engines or the main yard which was built to be manned by not less than six-footers.

Education at the Naval Academy consisted of trying to read the lessons assigned earlier in the day, but so many pages were assigned that it was usually impossible to finish. It certainly was impossible to find time to ponder over anything that seemed particularly interesting. In the classroom the instructors handed out questions and we wrote the answers on the blackboard. A mark was assigned for the work and that was all there was to it. There was no actual instruction. During my tour at the Naval Academy I met only two instructors who really taught. One of them gave me a grounding in navigation I never forgot and which proved to be invaluable.

Our course at the Naval Academy lasted only three and a half years because there was a shortage of officers and we were needed in the fleet. Nevertheless, we had to serve two years as "passed Midshipmen" before we could become Ensigns. The last six months at the Academy were spent in the first completed section of Bancroft Hall, palatial quarters compared to the wooden temporaries in which we had been living. The light in those temporaries where we studied was very poor and many of our textbooks were printed on shiny paper. This combination very likely was responsible for the defective vision of many of us in later life. Many of our textbooks were British—the first of many examples I shall give in other chapters of our unbelievably long dependence on British technology. It wasn't until late in the course that excellent books on marine engineering and naval boilers, written by some of the old engineer officers, appeared. The course was almost completed before we were furnished with handbooks, not regular textbooks, on electrical engineering and radio.

When I entered the Naval Academy it was obvious that my schooling was quite deficient from Naval Academy standards. Latin and Greek, English literature, mathematics through plane geometry and algebra with a smattering of physics and chemistry was very poor background for the Naval Academy. As a matter of fact, the head of the special prep school which I was supposed to attend for two weeks before entrance exams, predicted that I wouldn't get in. The reason for this prediction was easy to explain—I had attended his school only once to obtain copies of old entrance examinations, something I had never seen before. At that time I also became aware of "short cuts" in arithmetic, which I learned from a roommate, who, incidentally, failed on his exam. Instead of going to cram school, I devoted all my waking hours to looking up the answers to questions I did not know. That one attendance at the prep school was a fortunate incident. In my youth and ignorance I had thought

I would be examined on my knowledge of subjects. I learned then and there that examinations are nothing but a game of "questions and answers."

The discipline at the Naval Academy was severe by any standard. I never regretted it and I never objected to it. It was accompanied by cold justice. Our officers, from the Superintendent down through the Commandant, were officers and gentlemen. There were few we really disliked and there were many that we hoped we would grow up to be like. There is one thing this country needs more than anything else—discipline. I hope the Naval Academy will always lead in this field and not be led astray by the popular and sentimental notion that discipline inhibits originality.

Years afterwards I was to have a conversation with Dr. Harry E. Clifford, the well-known electrical engineer and educator, who was one of the organizers of the Navy Postgraduate School. He had just completed his membership on the Annual Board of Visitors of the U.S. Naval Academy. He said: "The Naval Academy is designed to educate naval officers, and that it does very well. There is one thing you people have there which is indispensable and I wish we could have it at Harvard. Your Midshipmen are under restraint during those formative undergraduate years. That is the finest thing in the world in connection with education."

I graduated from the Naval Academy in January 1905 and was assigned to an armored cruiser, the *Maryland*, then nearing completion. Like all passed Midshipmen, I performed the duties of a junior officer on deck and in the engine room. In 1906 I joined the *Dixie*, a merchantman converted for use in the Spanish-American War. She was known officially as an auxiliary cruiser of brig rig. To the uninitiated, she was just another "banana boat," one of a class of vessels stationed in the West Indies for interminable periods of time as a threat to ambitious revolutionaries. We carried Marines, and one night in 1906 put them ashore at Cienfuegos, Cuba, at the beginning of the Cuban Occupation. In the spring of 1907, having passed my final examinations as a Midshipman, I was commissioned an Ensign. In the fall of 1907 I was transferred to the battleship *Kansas*, which soon embarked on the first leg of the cruise around the world. On both the *Dixie* and *Kansas* I was a watch and gun division officer and did not perform any engineering duty. At San Francisco, in the spring of 1908, at my own request, I was transferred to the armored cruiser *Pennsylvania*, where I performed watch and gun division duties until I became the first assistant to the Chief Engineer.

In 1910, at my own request, I was transferred to the *Hopkins* as Executive Officer, and duty aboard that vessel turned out to be one of the best practical schools I had yet encountered. While I was unpacking my belongings on my first day on board there was a great deal of noise and confusion with a sound of the closing of water-tight doors. I stuck my head out of my stateroom and asked our one mess attendant, who was busily dogging down a water-tight door, what was the trouble. "Oh, another boiler tube has pulled out, and when that happens I always secure the wardroom." In response to my inquiry as to how often that happened, he replied, " 'Most every day." That epitomizes the situation. There was very little that was right about that destroyer. She had a bad history. Sometime earlier, while at sea off the Virginia Capes, her propeller struts had come loose and, swinging around on the propeller shafts, had wiped out the wardroom, and the vessel had had to be towed into the Norfolk Navy Yard with her decks awash aft.

It wasn't long before the Engineer Officer was transferred to the hospital and I became both Executive and acting Engineer Officer. My education began apace, as the Captain and I were the only officers on board.

The *Hopkins*, completed in 1903 and at that time 7 years old, was a destroyer 238 feet 9 inches long, 23½-foot beam, 6 feet mean draft, and highest trial speed of 29.02 knots, but no one ever saw her make it. She was equipped with a three-inch 50-caliber forward and another aft, and five six-pounders. On each side, in the waist, were twin-mount torpedo tubes. She was twin screw with two triple expansion reciprocating steam engines. The boilers were Thornycroft, a British design. A large steam drum was connected to two 14-inch mud drums by curved generating tubes, giving the boiler the appearance of a large "A." All three drums were connected in back by down-comers and a cross-connection. Of course the boilers were coal burners, and we operated, if we operated at all, under forced draft. She was rated at 10,000 I.H.P. A destroyer has always been and always will be a young officer's delight, but this one was different.

I soon learned that the forced draft blowers were unreliable and usually under repair. Forced lubrication had not come into use, and accessibility for repair and maintenance had never been thought of.

The crew was afraid of the boilers, and the Captain tried to solve that one by requesting me to visit the fire rooms often and remain long enough to give the fire-room gang confidence. That I did, but I always kept one eye on the escape ladder or on the entrance to the coal bunker.

The Babcock and Wilcox boiler-testing outfit for feed water had just been introduced into our large ships, so one day I asked the Chief Water Tender what was the condition of the water in a certain boiler. Obviously he had never even seen a testing outfit. He drew a cupful of water from the water gage petcock, took a draught, spat it out on the floor plates and confidently stated, "Sir, it's fresh."

Boiler trouble continued.

The Captain finally became so concerned about his boilers that he requested that the Bureau of Engineering have a Board of Survey inspect the machinery of the *Hopkins* in order to report on the safety of the boilers and recommend necessary repairs to the boilers and machinery.

About the same time he decided that the billet of Engineer Officer would not be filled and he assigned me to that job in addition to Executive Officer. At my request he delayed issuing me my new orders until the boilers had been inspected by the Board. This having been done, I proceeded to try to do both jobs.

Almost immediately after the inspection, the Captain ordered me to raise the steam pressure on the boilers, which is done by adjusting the setting of the safety (relief) valves. This I declined to do except under written orders on the grounds that he had previously requested a Board on these boilers, believing they were unsafe, and, while the Board had cleared them, there had been no material change since his original fears. If we did have an accident, he wouldn't have a leg to stand on.

The ship went on her way to do the usual things that destroyers were doing at San Pedro and San Diego, but I was soon ordered up to Mare Island for examination for promotion. During my exams I read in the paper that while underway several tubes had pulled out of the boilers of the *Hopkins* and that several men were killed and some scalded.

Before I rejoined the ship at San Pedro, the Board of Investigation on the accident had completed its work. I was not needed as a witness.

The *Hopkins* was in a bad way. The skipper was detached and for some time I was the only officer on board. Finally a passed Midshipman reported on board for duty. We had one good boiler, one was burnt badly and required retubing—a Navy Yard job—and the other two leaked too badly to raise steam in them. There was no Los Angeles harbor in those days, and we kept the one machine shop in San Pedro busy night and day turning out steel plugs for boiler tubes.

While at sea at various times, I have had occasion to observe the stability of ships. The most stable ships of course are the ones with the greatest metacentric height. The metacentric height is the vertical distance between the center of gravity and the center of buoyancy. The center of gravity is below the center of buoyancy. A ship with a large metacentric height is known as a "stiff" ship because she rolls rapidly and forcibly. Such a ship is uncomfortable for passengers and rolls too rapidly for gunfire, so a moderate metacentric height is desirable, although passenger vessels as a rule have much smaller metacentric heights than do warships.

When I received my first command, the *Hopkins*, one of the first things I did was to look at the stability curves, and

I was delighted to learn that she had a righting moment at 90°, or lying on her side. All I had to do on the *Hopkins* in respect to stability was to make her decks tight and keep her hatches tight.

I now had plenty of time to look around. Also I had to get three boilers tight for a trip to Mare Island where the ship was to be inspected by the Board of Inspection and Survey.

I have mentioned before that the three drums were connected at the back by down-comers and a cross-connection, perhaps four inches in diameter. My proposal to remove this cross-connection in order to see what the tube ends looked like in the back end of the mud drums was greeted with the information that the cross-connection had never been off, that the joints were ground and, if taken off, could never be put back by the ship's force. In other words, this cross-connection between the back ends of the mud drums had never been inspected. Nevertheless, I took it off and reached into the mud drum and, with my bare hands, tore off all the tube ends within reach. It was much easier to plug tubes with both ends of the mud drum open; otherwise it was necessary to tie a plug on a slice bar and work it arduously into the tube hole in the drum. Each end of a defective tube is plugged and, eventually, the tube is burned out. By the time I had made the boiler tight and the tubes had burned out, the steam drum, with the plugs sticking through, looked like a giant mangle.

Finally, having done all that was possible we got underway for Mare Island. We had almost reached the Golden Gate when things began to happen. Two boilers began leaking so badly that we had to haul fires. On the third boiler, leaks were bad enough so that we reduced pressure until it was just sufficient to run the auxiliaries and the steering engine. Again we started plugging tubes. The procedure was as follows: As soon as the water was dumped out of a boiler and while the steam drum was still steaming hot, we laid a plank inside at the bottom of the steam drum and a Chief Water Tender went in, prone, and plugged his end of the leaky tube. As soon as the Chief had plugged his end, which was the more difficult on account of heat and steam, he was rewarded by being allowed to drink all the whiskey he could drink without removing his lips from the bottle. There was no dearth of volunteers. Meanwhile, we had been drifting toward Pt. Montara, California, at the rate of three miles an hour. Finally we got two boilers tight, raised steam in them, and again we were on our way. Mare Island had sent a tug upon receiving our report of being disabled, but we made such good time that we met the tug halfway, at The Brothers, and waved her back to Mare Island.

The Board of Inspection and Survey was next, and they certainly told me what they thought of the ship and everyone who had ever been on it. When they began to criticize the ship's organization, I informed the senior member, a Rear Admiral, that he was going far beyond his precept. That was before selection for promotion. Two members of the Board who were strangers to me signalled me in a friendly fashion to pipe down and I took the hint. During the lunch hour they advised me, in spite of their admiration for my spunk, to let them handle any questions involving my responsibility and the authority of the Admiral under the precept and, for heaven's sake, not to tell the Admiral where to get off.

As a result of all this inspection, a huge amount of repair was recommended.

One afternoon, as I was returning to the ship, I learned that the aide to the Secretary of the Navy for Material had been to the Yard and had decided that the repairs were too expensive and that the ship should be scrapped. So, here was my first command, only a few weeks old, a command I had received as a junior to all the other commanding officers of destroyers, although it was true I received it because no one else wanted the job or would take it, but still a command.

I hopped the boat for San Francisco immediately and was fortunate enough to interview the aide in his room at the Fairmont Hotel. He was about to retire, but was very gracious about the whole affair. I persuaded him that I could cut the repair bill in half by doing a lot of work with the ship's force. The Admiral went along with me and I returned with the good news.

One fine day we were on our way again with a ship that was virtually new and with close fitting pistons and valves that should give us good speed and better economy. But my education in this floating university was not over. I had to learn more about engineering, something about accountability, and a lot about men.

For some time I had been trying to straighten out the ship's books and reconcile them with the inventory of equipment on board. It just couldn't be done. For instance, when a torpedo had been lost after firing, a form card had been filled out and submitted to the Bureau of Ordnance to report the loss. An entry was made in the inventory book, using this report card as authority to expend the torpedo as a loss. Of course that doesn't make sense. Material can be expended only after a report by a surveying officer or board who affixes the blame, if any, for the loss. The more I tried to reconcile "on hand" with "on charge," the worse things were found to be. Finally I embarked on a devious course. I wrote the Chief of the Bureau of Supplies and Accounts for permission to start a new set of books. This request was granted, but I probably read into the approval something which it had never intended to authorize, and I took up on the new books only equipment which was on inventory. Desperate affairs required desperate remedies. Even then, before we returned to the Navy Yard, I prepared a list of missing material. As soon as we were tied up to the sea wall, my scouts would go to the dump and procure articles similar to the ones missing. Then we were ready for the surveying officer.

The ship's mess was terrible. The men were poorly fed compared with other destroyers. The cooking was good, but there was not enough of it. Later we had reason to believe that the ship's cook was signing for more food than was being delivered—the old kick-back. At one time, when there was not enough food to go around, I was forced to issue an order that, before each meal, the food would be distributed plate by plate on tables below decks while the ship's force stood in ranks on deck until the distribution had been completed. My Chief Bosun's Mate, who was directly in charge of the crew, informed me that if I carried out this order there would be a mutiny. I told him that my two officers and I would be present at the initial formation inaugurating the new mess procedure and without side arms. If anyone wanted to start a mutiny, they would have a wonderful opportunity. Nothing happened except that I got rid of my Chief Bosun's Mate, and, with constant vigilance, the food was more equitably distributed.

One day a wrist pin bearing burned out in a connecting rod on a main engine because the connecting rod had not been "swung." The Machinist's Mate in charge of this engine room and I both learned something. A connecting rod has bearings on both ends which connect it to the cross head and to the crank pin or journal. Now the clearance in the bearing at either end of the connecting rod cannot be changed without "swinging." This means that the bearing must be tightened to running clearance, first one end and then the other and, in each case, the clearance must be satisfactory at the opposite end. Years later, when I was Fleet Engineer of the Battle Fleet, the Bureau asked me to revise the chapter in the Engineering Instructions on lining up main reciprocating engines "because you are one of the few left who knows how to line up a reciprocating main engine."

The *Hopkins* was soon shaking down into a real ship, and the next thing to do was to run full speed over the measured mile to see what we could do. She had not had a full power run for years. As a preliminary to the main event I opened her up in the Straits of San Juan de Fuca, having in mind the nearness of a chosen spot on the island of Vancouver where I could beach her handily if she got awash again from strut trouble. As we approached full power, the Engineer Officer reported shearing main bearing bolts and main engine holding down bolts, but the diagonal rod construction of the engine frame facilitated using wooden 4 × 4's to hold the engine together, and we finished our full power run. Soon after that I was ordered ashore, and the *Hopkins* ran her orthodox full power run over the measured mile. Upon docking her later, it was found that there were many cracked rivets in the struts. After the theory of torsional vibrations of propeller shafting was developed, we found out that the *Hopkins* had been ill-fated from birth. In the fall of 1911, having completed the customary six-year cruise, I was ordered to the Naval Academy. I did not wish to be an instructor and, at my own request, was ordered to industrial duty at the Portsmouth, New Hampshire, Navy Yard. Here I was outside machinery superintendent. The principal engineering job was designing and installing forced lubrication for the reciprocating main engines of armored cruisers. One of these, fortunately, was the *Tennessee*, and my experience with her was to help me a great deal when I became her Chief Engineer a couple of years later. At the same time, the first Navy Yard accounting system was being installed, and I had an opportunity to learn something about cost accounting which was to stand me in good stead in later years.

By this time I was convinced that my theoretical knowledge of engineering was far from complete, and I applied for a postgraduate course in engineering. Also, I had decided that I preferred engineering duty. Deck duty, watch standing, and gun drills bored me, especially drills. Fortunately, in 1912, I was able to join the first class of the new postgraduate school with one year at Annapolis and one year at Columbia University.

A group of brilliant educators, who assisted the authorities in setting up the course, were our instructors at the Naval Academy. Among them were Drs. Harry E. Clifford of Harvard, Herbert C. Sadler of Michigan, and Charles E. Lucke of Columbia. A young instructor from Harvard, Chester L. Dawes, immediately went to work and rebuilt the Midshipmen's electrical laboratory into something that bore no resemblance to the original. It became an instruction laboratory, not a museum.

Some of the Academy's courses previously set up by the civilian heads of physics, chemistry, and mathematics were terrible. I rebelled at spending days trying to determine the force of gravity at Annapolis when I needed to know more of thermodynamics and alternating current electricity. A Michigan professor took a calculus exam home for the university museum because it represented pure mathematics of no practical use and of a type not taught at universities, at least for engineers. We worked on it for 18 hours before completing it. We heard nothing about the use of calculus in engineering, although Clifford and Sadler were reeking with it. Metallurgy came under the chemistry department and, when my turn came to read an independently prepared paper on a subject I had chosen, I used words such as "eutectics" and other terms common in metallography. I was immediately interrupted by the head of the chemistry department, who wanted to know what a "eutectic" was and what was I "kidding about." There was a great deal of dry rot and stagnation in the Naval Academy of my Midshipman and postgraduate days. The civilian heads of departments, who should have been the educational leaders, had not kept up. The Academic Board apparently had no contact with the outside world, academic or otherwise. A lot of this was corrected by some of the distinguished engineers I have mentioned above and, finally, by naval officers who later were on duty as instructors and who had previously completed the postgraduate course. Inbreeding, a cause of dry rot, is not a naval monopoly. It is noticeable in universities and corporations as well as in government, and one must be constantly on guard against it.

Our instructors at Columbia, Professors Charles E. Lucke, Walter I. Schlicter, Morton Arendt, Frederick W. Hehr, and Walter Rautenstrauch were excellent. The Naval Academy was a memory course. Since I was taking mechanical engineering, most of my work was with Lucke, and he did not run a memory course. What a relief! Dr. Lucke greatly impressed me with his belief that the future of steam engineering depended on the employment of higher pressures and temperatures. In fact, the thesis I wrote for my degree was devoted to the thermodynamic advantage of reheating.

I was quite dissatisfied, however, with our curriculum at Columbia, and by the time the Christmas holidays came along I had decided to do something about it. The curriculum did not include steam turbine design and it did include the design of commercial elevators. The best book I could find on turbine design was published in England, still another example at this late date of our colonial ancestry. I studied this book from cover to cover during the Christmas holidays, at the end of which I demanded that steam turbine design be added to the curriculum and that elevators be dropped. I was threatened with being dismissed and sent to sea, but I had my flanks well protected because the engineers' post-graduate school was under the Bureau of Engineering, and I knew I would be supported by that Bureau if it came to a showdown. Fortunately good sense prevailed, and I got my course in turbine design.

At this point it is interesting to look back at some of the origins of technical education in the Navy.

As long ago as 1838 Captain Matthew Calbraith Perry (who later, as a Commodore, opened Japan to commerce) suggested to the Navy Department that a training school for naval engineering be established. This was seven years before the beginning of the Naval Academy. It is not surprising to learn that Perry was dead before his idea was adopted. The Engineer Corps was established in 1842. In 1863 the Engineer-in-Chief, Benjamin F. Isherwood, in a long and interesting letter captioned "Education for Engineers," wrote to the Secretary of the Navy: "The Naval service and the country have suffered injury that can scarcely be estimated from the want of thorough, capable and well educated engineers."[1] He recommended that naval engineers be educated at the Naval Academy. His views naturally excited considerable opposition. In 1864 Congress authorized the education of steam engineers and naval constructors at the Naval Academy. The Naval Construction Corps was established by Congress in 1866.

[1] *The Steam Navy of the United States*, Frank M. Bennett, Warren & Company, Pittsburgh, Pennsylvania, 1897, vol. 2, p. 655.

Nothing was done about education in naval architecture until 1879. In that year two cadet engineers of unbelievable foresight and vision, Richard Gatewood and Francis T. Bowles, saw a future in naval architecture and felt the need for special education in it. They were successful in obtaining for themselves a three-year course at the Royal Naval College, Greenwich, England. This was the beginning of the modern Construction Corps.

Other graduates of the Naval Academy each year were sent to Europe for architectural courses until about 1898 when a postgraduate course in that subject was established at the Naval Academy. Beginning in 1901, young officers selected for the Construction Corps were sent instead to Massachusetts Institute of Technology for their postgraduate work.

While the naval constructors were pursuing their forward-looking and enlightened course, what were the naval engineers doing? They were doing just the opposite.

The old Engineers Corps was abolished in 1899 by amalgamating it with the Line. All Midshipmen at the Naval Academy were required to take the same curriculum, which included a good course in operating but not design engineering. The old Engineer Corps killed itself because it had no vision and didn't reach out for postgraduate education. All this is strange because, until the rise of the central power plant, naval engineers were leaders in steam engineering. Possibly there was dead wood in the Corps, but the Corps had produced Ira N. Hollis of Worcester Polytech, Mortimer E. Cooley of Michigan, and William F. Durand of Leland Stanford, all pioneers in engineering education.

The period of "every naval officer his own engineer" got under way. Outstanding naval engineers like Robert S. Griffin, Charles W. Dyson, Emil Theiss and others refused to qualify as "straight" line officers and did yeoman's work in engineering in this interim period.

Engineering afloat or operating engineering became more and more in the hands of the Warrant Machinists, very able and practical men as a rule, but usually with no theoretical engineering education. The warrant officers also maintained a very effective lobby in Washington, and Congressional reaction to the lobby was noticeable. The warrant officers were carrying the engineering load and they naturally were seeking increased rank and increased remuneration. If it had not been for the provision for postgraduate instruction in engineering and later for engineering officers only, or "EDO's" as they were called, the warrant officers might easily have become a new Engineering Corps.

It is interesting to note that postgraduate education in the Navy was started, not by the Bureau of Naval Personnel (then called Navigation), but by the materiel bureaus. Eventually Personnel took control away from the bureaus.

The shifting of the control of an agency from the hands of the originator to a central activity through amalgamation or merger is an interesting phenomenon in organization and can be noticed also in corporations and in government. It has its advantages and its disadvantages. It introduces another link in the "chain of command" and does not permit full exercise of leadership by those who conceived the idea. It does facilitate the formulation of a general policy but at the sacrifice of sensitiveness. This passion for pooling the interests of subdivisions of an organization into a part of a central organization is something that should be watched carefully because it can lead to all the stupid evils we associate with the word bureaucracy.

There seems to be no means to prevent centralization, but we should at least be aware of its evils and remember that centralization puts another insulating layer between those who originate and perform and those who are policy makers.

In 1917 the first officers were selected for engineering duty only and they were usually postgraduates. They became extra numbers in the line. The term engineering duty only, or EDO, is not a particularly happy choice, appearing to represent animus or lack of acquaintance with the unusual opportunities presented by the English language for exact and gracious expression.

Those of us who joined the EDO's took a gamble. I had

made up my mind to devote my life to engineering, in the Navy if possible, outside the Navy if necessary. The law establishing our existence did not guarantee our future. A Chief of the Bureau of Engineering about this time expressed the Navy's doubt about the EDO's as follows: "The EDO," he said, "is like a mule, no pride of ancestry and no hope of posterity."

In 1940 the Bureau of Engineering and the Bureau of Construction and Repair were merged to form a new bureau, the Bureau of Ships. The Construction Corps was abolished and its members, the naval architects, became EDO's alongside the original EDO's. This new group also included the aeronautical EDO's.

The selection of young officers for specialization in naval architecture and engineering, before the merger, was approached from two distinct points of view. The constructors were selected for both postgraduate work and induction into the Construction Corps after graduation from the Naval Academy on the basis of academic standing. "The selections for this course (and for the Construction Corps) were in all cases from lists of volunteers, priority being given to class standing. In the first 20 years all of the appointees except 5 were from the first 5 members of the classes and during this period 16 were number 1 graduates of their classes."[2]

Candidates for the Construction Corps were therefore the result of one selection, i.e. class standing.[3]

The EDO, on the other hand, was the product of two selections. First he had to be selected for a postgraduate course in engineering on the basis of his accomplishments in engineering at sea and was usually about thirty years old. Second, the EDO was selected from postgraduates in engineering on the basis of what he had contributed to engineering ashore and afloat. It took the vote of every member of

[2] *History of the Construction Corps of the United States Navy*, U.S. Government Printing Office, Washington, D.C., 1937, p. 40.

[3] See Appendix J for the comment of Mr. William LeRoy Emmet in regard to selection on the basis of class standing.

the selection board to select, for EDO, a candidate who was not a postgraduate. In both selections academic standing was considered but not exclusively. The big question was: Is the candidate in production as an operating engineer? The results achieved by these entirely different methods of selecting naval constructors and engineers merit serious consideration.

On August 3, 1914, while I was doing practical work in the New York Navy Yard and still undergoing postgraduate education, I received telephone orders from Washington to report immediately on board the *Tennessee* as Engineering Officer. I reported that evening, and we sailed for Europe the following forenoon. World War I, European phase, had begun.

When we left the Port of New York on August 4, 1914, the *Tennessee* was not in materiel or personnel readiness to go to sea. She had been built and completed by William Cramp and Sons in 1906 and was at that time eight years old. In addition to all her other troubles, her machinery exhibited the acme of inaccessibility. It was two or three times as hard to do a job on the *Tennessee* as it was, for instance, on the *North Carolina*, virtually a sister ship built and completed at the Newport News Shipbuilding and Drydock Company hardly two years later. I am reminded of a few words from the well-known song, "The Armored Cruiser Squadron"—

> Oh why, oh why did Uncle Sam
> Build those ships, not worth a damn,
> In the Armored Cruiser Squadron?

There was a saying in the Navy of that period that the quickest way to learn engineering was to get ordered to a rundown ship as Chief Engineer. On that basis I was doing all right.

The *Tennessee* had been a receiving ship in reserve at the New York Navy Yard and, as her material condition became apparent, there remained in my mind an indictment of fleets and ships in reserve.

There was great activity on the *Tennessee,* what with putting aboard sea stores, coaling ship, receiving high-ranking officers from the Army and Navy, a delegation of bankers from New York and what was at that time the colossal sum of three million dollars in gold coin for use in helping American refugees in Europe.

We had no sooner turned over the main engines the first time when I began to find out what sort of craft I had inherited. As soon as we opened the main throttles, the relief valves lifted on the main condensers, so we immediately closed the main throttles and looked around to see what was the matter. The trouble was obvious. Many of the levers for the cylinder drains had been put on backwards so that, while the labels indicated that the drains were closed, actually half of them were open. Therefore, when we opened the main throttle, the pressure naturally built up rapidly in the main condenser.

We started off for Europe at the lively clip of 17 knots, and about the time our distinguished visitors and passengers were getting settled the one and only one-ton Allen dense-air ice machine went dead. This was a serious problem because about 13,000 pounds of meat, the only fresh meat we had on board, was endangered. The false plates which constituted the valve seats were rusted in, so new ones could not be installed.

I was forced to take for repairs the best mechanic I had on board who happened to be a warrant officer in charge of the starboard engine room. After twelve hours' work at night, using broken pieces of file, he was able to eliminate most of the scores of the valve seats and we could hear the happy chugging of our one refrigerator the following morning.

As we approached the mid-Atlantic, frequent inspection of the upper and lower coal bunkers had convinced me that the coal account had not been kept properly and we were actually about 500 tons short. I slowed the ship down to

10 knots, much to the dismay of everyone on board, including the skipper, one of the last of the "sundowners."[4]

The conversion of our fleet from coal to oil had proceeded so far that only a small part of our fire room force had had any experience in burning coal. Also, our crew had been hastily assembled by details from many other ships and represented the ragtail and bobtail of the Atlantic Fleet.

As the ship began to roll and pitch, all the auxiliary drains in the fire rooms became choked up from an accumulation of ashes in the bilges to such an extent that we could not pump bilges. In order to free the auxiliary drains, we had to bail out the water and ashes in the bilges into coal buckets, hoist the coal buckets on deck and then dump them over the side. After that we cleaned the bilges and restored the efficiency of the auxiliary drains, but not until we had used the main drain to pump out most of the water in the bilges, it being the practice—and good practice—never to use the main drain except in emergencies. This ship ejected ashes overboard by hydraulic means although, fortunately, the ash whips were still in commission. There was a sharp bend in the piping of the hydraulic ejection system in the upper bunkers and this bend was protected from erosion by steel plates which were replaceable as they became worn out. No one had worried about these erosion plates for so long that plate after plate was cut through and the containing bunker flooded. Work on the ejection system continued night and day until, finally, the system was in good working order.

Among the crew, assembled from so many different ships, no one knew anyone else. If a man did not report for his watch, his petty officer had no way of locating him. To correct this situation, I issued an oral order that, at the end of his watch, a man would be permitted to leave his station temporarily for the purpose of going up on deck

[4] Webster defines a "sundowner" as "a very strict Captain, originally one who compelled Midshipmen to return from shore leave at sundown." My own information is that he was a Captain who ran a "taut ship," sent down light yards every night at sundown, and was, in general, an S.O.B.

to bring down his relief. After a crop of black eyes appeared the situation improved. We had the hardest time rousing out the midwatch. They simply wouldn't stir, so I armed the masters-at-arms with baseball bats. These, lustily applied to the bottom of occupied hammocks, turned out the midwatch very successfully as well as everyone else sleeping on the gun deck. Of course I made the point of being personally present every night for this operation. One night the loud noise from the gun deck where the men were berthed reached the ears of the Captain on the bridge. He was as hard and regulation as they come. He sent for me and inquired as to whether or not I was doing anything in violation of the Navy Regulations. I told him if he wanted to reach England he better leave the Navy Regulations to me. He wanted to reach England.

But the worst was yet to happen. Early one morning I was called by the Officer of the Watch, who informed me that he had heard a loud click in the starboard main engine and then immediately the receiver gauges of the high pressure cylinder and the intermediate cylinder had equalized. It was obvious that the engine had compounded itself, and we were forced to run on one engine while we disabled the other for repairs. An inspection of the high pressure valve chest found the valve rings all broken and there were no replacements on board!

There was always a controversy in the Navy, which I don't think was ever settled, as to whether split rings or solid rings were best for this duty. I, personally, preferred solid rings. But, to continue the story, we took out the broken parts, removed the high pressure valve altogether and soon were running on one good engine triple expansion and the other engine compounded.

As we neared our destination and the factor of safety with respect to our arrival increased, I raised the speed by increments until we entered Falmouth Harbor at 17 knots.

After procuring new valve rings in Falmouth, we were able to operate both engines at triple expansion.

Later on, when we were in the Eastern Mediterranean,

we broke a high pressure valve chest liner so badly that we had to obtain a rough casting all the way from the New York Navy Yard. This we machined on board and put in as a forced fit, but before we forced it in we covered all of the binding surfaces with saleratus. We were sick with fear one minute that the valve chest would not be tight enough and would work out and the next minute that it would be too tight and split the engine. But we fitted it as tight as we dared and it stayed there. By this time I realized that I had a bear by the tail. This was back in the days of the Babcock and Wilcox longitudinal boiler with vertical headers, cross-connection boxes, and side-box water walls. The use of boiler compound was in its infancy, and so much had been used in the past that the hand-hole gaskets were eaten away and were continually letting go. We renewed hand-hole gaskets all the way across the Atlantic, which meant hauling fires and relighting them again no less than 32 times.

The bunkers and double bottoms were in such bad condition that I sent back to New York for hundreds of gallons of red lead, as many pneumatic scaling hammers as they would allow me, and hundreds of yards of hose. I had a hard time getting the Captain to give me permission to work three shifts with special daylight liberty for the night shift. As long as he had been in the Navy he had never heard of such a thing. The clincher was—did he want to pass successfully the next inspection by the Board of Inspection and Survey? He did! By working three shifts for months we were able finally to get the ship in such a condition that, one year after our departure for Europe, we were able just to squeak by the Board of Inspection and Survey.

Those were the days before the theory of torsional vibrations was known, and the breaking of propeller shafts was a frequent and disastrous occurrence in the Navy. In torsional vibrations the propeller acts like a flywheel, alternately storing and releasing energy. This causes long propeller shafting to twist and untwist and, ultimately, to crystallize and fail. This failure may be accelerated if the steel of the shafting contains graphite or slag inclusions. Those were also the days when the Navy, which had done so much pioneering for metallography, was struggling hard with the steel companies to get them to agree to include metallographic examination in the specifications. We were thinking especially in terms of shafting.

Every time we were under way, sooner or later the ship began to vibrate. The Captain immediately sent for me, which generally meant a climb of one hundred feet from the foot of the main condenser to the navigation bridge, to tell me to be sure that I kept the engines out of step. It was the pounding of the whistle on the stack which always made the Captain think the engines were in step. I kept them out of step but, unfortunately, the ship vibrated badly at times anyhow.

We left the Navy Yard at Philadelphia, Pennsylvania, to intervene in the Haitian Rebellion, and the whistle did not vibrate on the entire trip to Port au Prince. The Captain congratulated me on having taken the vibrations out of the ship. He did not know that, while we were at the Yard, I had taken the precaution of having the whistle welded instead of bolted to the stack.

Before we left the Mediterranean in August 1915, and before the Gallipoli Campaign, we wrote the last episode in the history of ultimatums delivered by a "senior naval officer present." Our Captain, Benton C. Decker, had sought and obtained permission to visit Smyrna, Turkey, and to make an official call on the Bey of Smyrna. He was ordered to anchor at Vourla Bay and, upon arrival there, was told by the Turks that they would furnish him with land transportation to Smyrna. This was unacceptable to our skipper and he proceeded forthwith to Smyrna in one of our steam launches on November 16, 1914.

He had not proceeded far before we heard firing and we could see from the splashes that the Turks were firing at the Captain's boat. The second in command, Lieutenant Commander Earl P. Jessup, promptly called all hands, manned the battery, bore-sighted guns and made all preparations to get under way. He also sent a message—an ultimatum—to the Bey of Smyrna to the effect that, if he did not desist his firing on our Captain's boat, he would get under way and bombard the city of Smyrna.

However, before this happened, the Captain returned to the ship somewhat crestfallen. All the shots at his boat were well-placed misses, and we came to the conclusion that the Germans had already infiltrated the Turkish army. Also, though we had heard the scuttlebutt that the mine field protecting Smyrna Harbor was no good, we felt it might be a wish instead of a fact.

We got under way all right and that night, by order of the Ambassador to Turkey, Mr. Morgenthau, at whose disposal the *Tennessee* and *North Carolina* had been placed, we headed not for Smyrna but for the Greek island of Chios.

The wardroom was furious, and one of our members reported the incident to the Associated Press correspondent on that island. The correspondent didn't believe the story and refused to transmit the information unless he was furnished the cost of cabling to New York. A collection was taken in the wardroom and the cable was sent. The receipt of the news created a sensation in the United States and no one except the wardroom ever knew how the news leaked out. The Captain just couldn't understand it.

Not long afterwards we received by a cable a general order which prohibited any senior officer present from issuing an ultimatum without first communicating with the Navy Department.

As I said earlier, Captain Decker was the last of the "sundowners." I was his third Chief Engineer. The first chief resigned from the Navy; the second chief went to the hospital. I decided I would have to act a part, to be harder than he was, and never to let him get anything on me. I became a severe disciplinarian; I made my "black gang"—the engineers—the best-dressed division on board, and I never ceased looking out for my men ashore or afloat. Because they knew I was looking out for their best interests always and because they knew I fought for them at the drop of the hat, they gave me all they had. I never worked men any harder in my life. I had to keep the machinery going. I cannot remember ever bringing a man to "mast" for punishment by the Captain for infractions of regulations. I punished them myself and in my own way. It is true that once one of my men reported me to Secretary Josephus Daniels for being a Prussian, but the letter was sidetracked in the Navy Department and nothing ever came of it.

Before I left the *Tennessee* I had a showdown with the Captain. No one, I am sure, had a more loyal and hard-working assistant than he, but such things happen on long cruises. He used to tell me that when we went into battle I, his Chief Engineer, should look out for everything below decks and he would look out for everything above decks—there was no place for the Executive Officer in his battle organization.

Josephus Daniels, then Secretary of the Navy, was pushing a pet plan for schools aboard ship. So we had to have schools on the *Tennessee* and the men hated it. Some of our best mechanics who had signified their intention to ship over when their enlistments expired changed their minds. They had left schools of their own volition at an early age, and they had no intention of staying in a Navy that sent them back to school, worst of all, aboard ship. My machine shop was run by a fine machinist who could never catch up with his backlog of work no matter how long he worked. He was well satisfied with his job and his ship until school came along, but he was an unwilling pupil while the work was piling up in his shop. I suspected the Captain, in his phenomenal zeal for school, of putting the school ahead of the efficiency of his command, of playing to the galleries. Be that as it may, I was outraged that anything so asinine should interfere with my preparations for the big inspection by the Board of Inspection and Survey as soon as we got to New York, and I embodied all my belligerent sentiments in a letter to the Captain.

The next morning at eight o'clock, when the notes of the "Star Spangled Banner" hardly had died out, the Captain's

Marine orderly was knocking on my stateroom door and uttering those fateful words, "Sir, the Captain wishes to see you."

The Captain was highly displeased that I had written such a letter to him—the contents would be known all over the ship to the detriment of the ship's force and the school system—and it would have been so much better if I had told him privately about my opinions of school and never had written that letter.

It looked like I would be put "under hatches" or suspended from duty, so I decided to beat him to the draw. I told him that obviously he was intending to put me under suspension and that he couldn't do it. Captain Decker at least had a good mind and he knew that I wouldn't question his authority to suspend me whether he was right or wrong, so instead of flying into a rage he merely asked me why not.

I told him that he couldn't afford to put me under suspension because the crew knew me to be the staunchest supporter he had and, if he punished me, the crew would say, "So what? If he punished his best supporter, what can the rest of us expect?"

It is, I think, to the credit of the late Captain (later Rear Admiral) Benton C. Decker that he could and did change his mind when circumstances required. We parted and were always good friends.

A ship of the vintage of the *Tennessee* was planned and built without any thought whatsoever about space required for the administration of an engineer's department. The typical cramped, ill-ventilated, dingy engineer's log room was located on the berth deck. We had to have a place to test boiler and feed water for salt and alkalinity or acidity with a standard Babcock and Wilcox boiler testing outfit consisting of bottles of silver nitrate, test tubes, pipettes, litmus paper, etc. In desperation we located our testing in a boiler uptake on top of the armored grating—a cramped location and usually hot as Tophet.

During one period of our cruise in the Mediterranean we were bothered by appreciable salt in our boilers, not too much for drinking purposes, but too much for boilers. We couldn't run down the location of the salt-water leak. The fireman who tested boiler water was an old-timer, so apparently there couldn't be anything wrong with the test. The salt would appear first in one boiler, then another, sometimes in the feed tank, sometimes in a water bottom, seldom twice in the same place, and without any systematic appearance.

In desperation I decided one day to watch the man who couldn't make a mistake testing water. We were both sitting on cracker boxes in the hot and humid atmosphere of the uptake. We calibrated and tested when, lo and behold, I saw a drop of sweat from my fireman's forehead fall into a sample of the boiler water he was testing. We had our answer to the phantom salt leak.

Some of my best mechanics always got hopelessly drunk whenever they went ashore on liberty. I handled that situation by telling them that they must not go ashore without my permission, even if their names were on the liberty list signed by the Captain. I did let them go ashore, however, if they asked my permission and told me what member of the Engineer's Force had volunteered to bring them back to the ship.

One of my very best men, and one whom I subjected to this surveillance, was C. Locke, Jr., a Machinist's Mate first class, and, because of his mechanical knowledge and ability, I put him in charge of the starboard engine room, the lead or senior engine room. No matter how many times I tried, the Bureau of Navigation (now Bureau of Naval Personnel) would not make him a Chief Machinist's Mate. Finally I had to make a trip to Washington on other matters, and I took the case up with a young officer who handled promotions of enlisted men for the Bureau. I demanded to know why my candidate couldn't get promoted. I was shown his service record which contained several "red" spots—drunkenness—and was told that that was the reason. Here I was, a Chief Engineer, trying to make an indifferent old coal burner operate and the Bureau was telling me what they would do and wouldn't do for my men. I said as much to the young officer and more.

I told him that I couldn't operate my ship with Sunday School teachers and the like, that my Machinist's Mate was a real man in spite of his record, and I demanded his promotion. He was promoted. Unfortunately and unknowingly I probably did the worst thing for him that I could have done. Months afterward, and after I had been detached, the *Tennessee* (renamed *Memphis*) was blown ashore by an unprecedented West Indies hurricane near Santo Domingo city, unprecedented because the north side and not the south side of Santo Domingo is ordinarily subject to hurricanes. My new and proud Chief Machinist's Mate was killed. I am sure he was at the throttle of the starboard main engine when the bulkheads caved in; he was that kind of man.

In the autumn of 1915, and before the loss of the *Memphis*, I was transferred to the armored cruiser *Pittsburgh* (formerly the *Pennsylvania*) as Chief Engineer and Fleet Engineer of the Reserve Pacific Fleet, Rear Admiral William F. Fulham in command. As long as we were based at the Puget Sound Navy Yard the routine was interrupted only by occasional trips at sea to test an individual ship. In 1916 events moved rapidly. On March 9, Pancho Villa raided Columbus, New Mexico, and General Pershing entered Mexico on March 15. The Reserve Fleet kept its identity as such, although all ships were activated and on duty in Mexican waters. Early in 1917 the Pacific Fleet absorbed the Reserve Fleet, and I became Fleet Engineer of that Fleet on the staff of Admiral William B. Caperton. We were on our way south before war was declared on April 6, 1917, and, after passage through the Panama Canal, learned that our destination was Bahia, Brazil, to establish a Fleet Base.

Upon arrival the Admiral learned that South America was in a political ferment. There were rumors that the Germans intended to set up a new country with the object of incorporating all German settlements in Argentina, Chile, and Brazil at least as far north as São Paulo. Uruguay would be included for geographical and military reasons. Brazil and Uruguay immediately requested that the Admiral visit their capitals with the Fleet.

The Pacific Fleet had been assigned to the South Atlantic to protect grain shipments to Europe, but the Admiral felt that the political situation in South America was more important. He felt that the best place for a South Atlantic Fleet Base was the harbor of Rio de Janeiro, so he proceeded to Rio for that purpose and so reported to the Navy Department. The reception of the Fleet in that port indicated that he had gauged correctly the apprehensive view of public opinion. Finally the diplomatic pressure was so strong that visits were made to Montevideo and Buenos Aires.

Argentina had preferred that the Fleet visit Bahia Blanca, an ocean port with a naval base, but the Admiral insisted on Buenos Aires. I think Argentina was sincere in wishing to avoid the possibility of an incident due to a large German population. As Fleet Engineer, I was not enthusiastic about a trip up the Rio de la Plata. The local offshore wind, the "pampero," had been known to blow so much water out of the river that ships grounded in mid-channel. This was the pampero season. Several of the commanding officers talked to me apprehensively about the trip, not just about grounding but also about the danger of the main injection to the main condensers becoming clogged with mud from the river bottom. However, there was a way out of the latter contingency. With those cruisers, it was possible to operate both main condensers with both main injections closed and using the overboard discharge (a high opening on the ship's side) on one side as an injection, the same circulating water thus passing in series through each main condenser and overboard through the overboard discharge on the other side. The circulation water passed from one condenser to the other through the main drain.

The trip to Buenos Aires and return was made without incident in spite of the fact that the ships drew more water than there was in the channel, a possibility because of the very soft, oozy bottom. They were said to have been the heaviest ships that ever visited that port.

At a private shipyard in Rio harbor, we maintained a coal pile and a stock of indispensable engineering supplies for the use of our cruisers patrolling the South Atlantic steamship lanes.

South America was all right, but I wanted duty in the North Atlantic, and, just before the end of 1918, I got my orders and took passage on the Lamport and Holt liner, *Vestris*, from Rio de Janeiro to New York. I didn't like the stability of the *Vestris* at all. With the slightest puff of wind she would lie over on one side and stay there for hours. I was not surprised when she later sank off the Virginia Capes with great loss of life. Her lack of stability and faulty loading contributed to the disaster. Curiously enough, if I had not made this passage on the ill-fated *Vestris,* I would have sailed on the Navy collier *Cyclops*, which left Rio shortly afterward with a cargo of manganese ore and was never heard from again.

Upon arrival in New York I reported on board the new battleship *Arizona* as Chief Engineer. She was new all right, but absolutely unreliable and required a lot of engineering attention.

The *Arizona* had a record for hardly ever going to sea without stripping her turbine blades. She was a direct-connected job with cruising turbines driving new-fangled reduction gears of the Melville-McAlpine floating frame type. There was only one man on board who could keep them in operation. He was a temporary Lieutenant and, in his own right, an excellent toolmaker. He understood lining up and repairing delicate machinery better than anyone else.

There had always been a tradition on the ship that she must never be backed full speed for danger of wrecking the turbines. One time, after we had anchored in the York River, Virginia, Naval Base, known in World War I as Base #2, the Captain sent for me and apologized for having backed the turbines full speed and expressed his amazement that I had not protested. I told him that everyone knew the *Arizona* was no good and the sooner the turbines were completely busted up and some new ones put in the better it would be for everyone.

In the meantime I started studying the blueprints in detail, and I went over all the reports of accidents in the log to try to find out what was the matter with the turbines and what must be done to make them reliable. The low pressure turbines were so large that when they were open for examination I frequently went inside the casing and, standing up inside the rotor, was jacked around with the rotor in order to participate in that favorite sport on the *Arizona*—listening for rubbing turbine blades. The temperature of the circulating water at that time of year was 31°F in the York River and not much warmer in the North Atlantic. Whenever I endeavored to take advantage of low vacuum the rubbing began. I decided the best thing to do would be to take .020 inch off all the low pressure rotor and casing blading. The Bureau of Engineering didn't think much of the idea but gave me permission. I was aided in this task by a detail from the New York Navy Yard and they did a fine job. Then an amazing thing happened. Increasing tip clearance results in decreased efficiency, but now we were able to obtain so much more vacuum without the blades rubbing that the overall result was a marked improvement in fuel consumption. After I was detached, and at the next visit to the Navy Yard, the tip clearance was increased similarly on all turbines and the *Arizona* no longer was prevented from making a trip to Europe. These turbines had a million blades, some of which were of arm's length. The low pressure turbine casings were so huge that the after end moved back and forth perhaps an inch with respect to the rotor and shaft due to temperature changes. When the upper casing was lifted but supported, there was no tip clearance between the rotor blades and the lower casing.

These monstrous turbines represented the extreme low in the development of the direct-connected Parsons (English) turbines and the extreme low in coupled cruising turbines with reduction gears. They should have been put in a museum to show young engineers what turbines looked like before the development of the Curtis (American) turbine and before we had learned how to cut reduction gears.

Before I left the *Arizona* an incident occurred which tells more than words about the position of engineering in the Navy at that time.

The Commander-in-Chief of the Atlantic Fleet was not satisfied that he knew what the condition of the flagship's machinery was. Also, he was not sure that the flagship's Chief Engineer knew either. So the Chief-of-Staff arranged for me to inspect the engineering plant of the flagship! Now this is a job which ordinarily falls to the Fleet Engineer, but there was no Fleet Engineer and no one on the Admiral's staff wanted the job because of the fear of getting too much engineering duty on their records, which they felt would militate against future good assignments and promotion. I was selected to make the inspection because of the success I had had in straightening out the *Arizona.* As a matter of fact, I found the flagship in excellent condition.

In the summer of 1918 I was detached from the *Arizona* and ordered to the Navy Department to the Office of Gunnery Exercises and Engineering Performances, later known as Fleet Training. Captain William D. Leahy (later Admiral) was ordered ashore at the same time to head the office. During World War I the submission of engineering performance reports had been discontinued, so now, when it was desired to lay down the characteristics of new vessels, no recent steaming data were available; hence the intention to restore to this office its former authority and responsibility. At the end of a year I felt that the purpose had been achieved and, at my own request, I was transferred to the Mare Island Navy Yard as Shop Superintendent. Before the transfer was effected, however, I was called on the carpet by the Chief of the Bureau of Navigation (now Naval Personnel) and accused of using "political influence" in regard to this transfer. After discussing the matter with the congressman concerned, I decided that "influence" is only "political" when the other fellow uses it.

It was a very interesting experience at Mare Island to install the first balancing machines. They were made by Akimoff at Philadelphia and probably were the first machines to effect dynamic as well as static balance. The amount and the location of the corrective weight was worked out by the theorem of moments after the experimental data on unbalance were obtained.

At Mare Island I learned a lot more about reduction gears. I received an order to hob the gears for a Naval auxiliary vessel which was building in the Puget Sound Navy Yard. We had one gear hobbing machine which had been built in England, and it had evidently seen hard service before the Navy acquired it. Upon examination and test it was obvious that it required a complete overhaul. The master mechanic of our machine ship, Mr. James Gee, and his toolmaker were extraordinarily competent and precise mechanics.

When I asked the Bureau of Engineering for an allotment of money to rebuild the English gear hobber, the Bureau refused on the grounds that the machine had never been used, that the overhaul was unnecessary, and unless I would agree to cut the gears on that machine as it stood, the job would be transferred to another yard. I had seen what kind of thread that gear cut on a blank (a cast iron model of the bull gear wheel) and I was positive that I was right, so I would not yield. Therefore, the work was transferred to another yard and cut on a similar gear hobbing machine; I learned later that the job was a failure.

When I became Engineer Officer of the Mare Island Navy Yard my problems became somewhat different. I had long wanted to clean up the foundry. It had a wooden roof supported by a complicated system of rafters that were characteristic of the Civil War period or perhaps even earlier. The interior of the roof and rafters were black as night with the accumulation of the dirt and dust of ages piled high.

When I broached the subject of cleaning up the mess to my master molder, H. G. Bailey, I was surprised to find out that this wise, quiet, affable and cooperative gentleman didn't want to go along with me. He didn't think the molders would like it. I was adamant. Over the weekend I sent men aloft with pneumatic lances to blow down all the dirt and dust. This was quite a job. Then I had the whole of the interior of the roof and all of the rafters sprayed with whitewash. That was anathema. No molder had ever heard of anyone's having the temerity to use whitewash in a foundry. It just wasn't done. But the place looked better and more cheerful, and I was satisfied. Later on Bailey admitted that the idea was not too bad because now the molders could see much further into their molds than they ever could before.

I received a rude introduction to finances one year when the Naval Appropriations Bill passed Congress and became law several weeks after the beginning of the fiscal year. The cut in new construction money was so severe that already I had overspent my monthly allotments. I couldn't lay men off as rapidly as I wanted to because they usually had accumulated leave to be taken and paid for. I felt as if I were headed for San Quentin. After I got that one straightened out, I decided to find some method of rough and ready accounting so that I would always be in charge of the financial situation. Returns from the accounting office usually were submitted three to five days—or more—after the event and that wasn't fast enough. Finally I worked out a conversion relationship between man hours and money, which was a great help, and later I developed the idea into a budget system at another yard.

In July 1922 my three-year assignment was completed and I went to sea as Assistant Fleet Engineer on the staff of the Commander-in-Chief of the Battle Fleet, Admiral William E. Eberle. The Battle Fleet was our main fleet and was stationed in the Pacific.

The flagship was the *California* and I had participated in the final stages of building that ship at the Mare Island Navy Yard. During the first year the usual cruises were made to the Puget Sound area in the summer and to Panama in the winter for problems with the entire Fleet, Atlantic and Pacific, and also with the Army. At the end of the first year, Admiral S. S. Robison relieved Admiral Eberle, and I became the Fleet Engineer of the Battle Fleet in 1923. The Fleet continued to base in the San Pedro-San Diego area, with a cruise to Australia and New Zealand in addition to the Puget Sound and Panama employment. The Australian cruise demonstrated that the Fleet had progressed greatly in reliability and self-maintenance.

In 1925 I became Fleet Materiel Officer on the staff of Admiral S. S. Robison who became Commander-in-Chief of the U.S. Fleet. The armored cruiser *Seattle,* formerly the *Washington,* was the flagship. I had not been on board the *Seattle* long before I noticed her peculiar sea-going characteristics. She would take a list with the slightest puff of wind or the smallest amount of helm. However, no one else seemed to think anything of it, not even the Fleet Naval Architect. The Columbia River from the sea to Portland, Oregon, is full of sharp turns and, in making this passage, the *Seattle* heeled over on every turn and reluctantly returned to even keel. Again it bothered no one but me. I was convinced that something had affected the stability of the ship and went to work to find out what it could be. I began looking for changes in the original plans of the ship and it wasn't long before I found what I wanted. The engineer's force some time or another had connected the wing (side) double bottoms to the feed water storage system. If they had kept these bottoms or tanks full it would have been all right, but they didn't. With all that "free surface" in their partially-filled condition, the water would rush toward the low side on each roll and increase the period of roll by making the ship roll deeper and return to vertical more slowly. I took the necessary action to keep these reserve feed tanks either full or empty and after that the *Seattle* behaved like a ship.

The *Seattle* was my last cruise. I went to the Puget Sound Navy Yard as Production Manager, and the Commander-in-Chief of the U.S. Fleet, Admiral S. S. Robison, went as Commandant. The Manager was Captain S. M. Robinson, later Rear Admiral and Chief of the Bureau of Engineering.

The Puget Sound Navy Yard was a mess; for years my jobs seemed to have been to straighten out messes. Later, when Secretary James V. Forrestal used me for the same purpose, he called me his "one-man task force."

The Yard had lost its ability to complete ships' repairs

on time or within its own estimates of cost. The Fleet was hollering for its ships which couldn't get away from the Yard, and the Yard was hollering to Washington for more money. The Yard was suffering an organizational breakdown.

The first thing I did upon arrival was to give the Yard an old-fashioned spring housecleaning. Every article in the shops or elsewhere which was not of immediate use was sent to the scrap pile if it were metal or to the dump to be burned if it were wood. All summer long the flames from the dump illuminated the Yard at night. Old pilot cabins were strewn all over the piers and dockside. Every time I burned a cabin, I saved the services of two men and I turned in two telephones. At the same time I discharged employees by the hundreds.

Ever since I had left Mare Island I had been thinking about developing some scheme to obtain better control of labor and cost through accounting. I felt that what I had accomplished at Mare Island in this respect could be developed much further. My ambition was to know each noon the basic accounting facts for that day. The accounting system reports came too late to have any practical value for the present even if they were excellent history. I wanted to know each noon where I stood on my expenditures and allotments. I refined my "conversion" factors with respect to cost of labor, material, labor overhead (shop expense), material overhead (shop expense), department overhead, general expense, cost of leave and holiday, etc.

I knew how much money I had available for expenditure on labor for repairs and alterations in a month by adding together all my allotments from the Fleet, the Bureaus, and other places. Then I divided this sum by my conversion factor. This gave me man hours. Dividing man hours by eight gave me the number of men I could hire and pay for per month and still come out even. But I didn't hire that number, I hired 90 per cent of it. There is a practical as well as a psychological reason for this. It saves money, obviously, but more than that. Men always fear working themselves out of a job. With my system that is not possible, and when they have convinced themselves that it is not possible their productivity rises.

I revamped the Schedule and Progress Section and had each supervisor telephone to the section the first thing in the morning how many men were working in his group. I checked the total against the number supposed to be working. Using conversion factors, I figured the charges for the day against each job order. Finally, I obtained the estimated charge for the day against each allotment. After I got this system fully organized, I turned it over to my Schedule and Progress Section and, by noon each day, the section presented me with a daily form, filled out, which showed expenditures by job orders for the day, and the percentage of completion of jobs.

The head of the Scheduling and Progress Section was Mr. Ralph W. Clark. He rapidly developed an astounding familiarity with shop and Navy Yard accounting, and it wasn't long before I was able to turn over to him all the work involved with our daily forms of accounting and progress.

Between the burning and scrapping of everything I could lay my hands on, together with the drastic payroll cuts, it was rumored around the Yard that I had been sent to Bremerton to close up the Yard and ship the remains to Mare Island. That was only a passing phase, however, in my relationship with my men.

Having cleaned up the Yard and shops and made the Yard solvent, I proceeded to budget all expenses in the Production Department, including the cost of repairs and new construction. I had never seen or heard of this being done before, but I was sure it would work and save more money.

I started with the "inside machine" shop because if my budget worked there it would work anywhere. Another reason was personal. Forbes Smith, the master mechanic of the shop, was the best marine engine man I had ever known. He was the most influential man in the Yard. I had known him ever since he used to swing a sledge in the crank pits of the main engines of the old armored cruisers. If I could sell my stuff to him, I could sell it anywhere.

I shall never forget the time I called him up to my office to help me prepare next month's budget for his shop. He didn't know how many crane operators he had nor how many sweepers, shop storemen, how many men there were in the central tool grinding room, how much money he needed for loose and hand tools, for repairs to machines, or cost of leave and holiday. In fact, Smith didn't have the slightest idea about the cost of anything in his shop. He didn't know anything about cost because no one had ever talked to him about cost before, but he learned quickly.

In no time at all the machine shop was saving money, and Smith said to me one day, "What on earth did we do with all that money we used to spend; I'm still giving the shop everything it needs."

From the machine shop I extended the budget to all shops. I even persuaded our power plant to adopt an annual overhaul schedule in order to avoid extreme fluctuation in the rates for light, power, heat, and compressed air.

I set the shop expense for a month, on the 20th day of the preceding month. Finally I set the Yard overhead for the following year. In all of this work a very remarkable and able man was of inestimable value to me, Mr. George A. Hastings, Assistant Shop Superintendent.

Budgeting paid big dividends when we constructed the cruiser *Louisville*. She was built for ten per cent less cost than a sister ship built in an East Coast Navy Yard.

When I joined this Yard, it had the highest overhead of any Navy Yard. My budget soon reduced this to be the lowest overhead of any Navy Yard, and my system of never hiring enough men to spend all my money cut down the cost of repair work. As for new construction, the *Louisville* speaks for itself. Later I was told that my budgeting of the Puget Sound Navy Yard preceded similar work in important parts of American industry by eight or nine years.

To make a record of my experience I wrote an article on the "art of budgeting in Navy Yards." The Naval Institute refused to publish it on the grounds that they did not handle such material, but it was published by the *Journal of the American Society of Naval Engineers.*

I submitted a report on my budget work to the Navy Yard Division of the Navy Department, but they neither acknowledged receipt of my contribution nor did they pay any attention to it. The Navy Department was simply not cost conscious and therefore not interested. In fact, no one paid any attention to my work on the organization of Navy Yards, but for me it was an invaluable experience.

In building the *Louisville,* I tested all the principal equipment in the shops before it was installed in the ship. It is true that all this material had been tested and passed by a Naval Inspector of Machinery in some Inspection District, but I did not accept that inspection as final or conclusive. The place to overhaul new machinery is in a machine shop and not aboard ship. That is too expensive, too difficult, and too time consuming.

Small corrections were made on some equipment, which did not excite much comment; but when I rejected the main reduction gears of the *Louisville* and refused to install them unless the contacts between the threads of the pinions and bull gears were reground, I started something.

I set up the gears in the machine shop, back to back—that is, with the propeller shaft couplings of each gear coupled together—and applied full load torque after the contact surfaces had been coated with Prussian blue. The lack of even and continuous contact was convincing.

I was told that the gears were all right because they had been passed by an Inspector. My answer was to ask them to send someone out to look at them. They did. The manufacturer of the gears sent the foreman of the machine shop in which the gears had been cut. He admitted a defect in the gears after seeing the results of the Prussian blue test. As a result the gears were lapped in, a laborious process, and the manufacturer paid the bill.

I had an experience with the galvanizing shop that made a deep impression on me. The galvanizing room had been operating for years without any accidents and suddenly, without any warning, there was a death. There were ugly rumors that it might be a case of murder. The gang all ate in the same room and there were suspicions that someone had put poison in the food of the deceased. Later I went down to Mare Island and checked on their galvanizing shop but found out that they never had any such fatality in their long experience. Not long afterwards the same type of accident occurred again. I don't remember whether it was another death, but it was at least serious illness. Why had this shop suddenly developed this characteristic after long years of operation?

There is one thing certain in this world; namely, that many people know about things they never tell you for the simple reason that they think the information they have is not pertinent. I learned accidentally that, originally, this shop had operated for years without anything except natural ventilation. Then, sometime before I arrived at the Yard, a doctor on inspection had recommended that blowers be installed to improve the ventilation. That was the trouble.

In galvanizing a deadly gas is generated, arseniureted hydrogen, which is about two and a half times heavier than air. Without artificial ventilation, this gas fell to the floor by its heavier weight and didn't bother anyone. When blowers were installed, with vents in the ceiling, the artificial ventilation mixed this deadly gas with the air at breathing level. As a result of my analysis and conclusions, suction blowers were installed with suctions in the floor, which is industrial practice where deadly heavy gases are generated in industrial processes. Medicine had entered industry, and we learned by bitter experience what we should have been told.

We balanced statically and dynamically everything that went into the *Louisville* that was susceptible to balance. There had been a great deal of progress in the science of balancing since I installed the first balancing machines at the Mare Island Navy Yard.

I had all my shops and offices gone over with a lux meter

and had all illumination brought up to the standards of candle power or lumens recommended. This is especially desirable around machine tools.

On one occasion I used dynamite aboard ship to save money and expedite work. The battleship *Colorado* came in one day with a bow full of concrete. She had been in a collision and concrete had been used lavishly to make the bow tight for the sea trip from San Pedro to Bremerton.

For at least two days we tried pneumatic drilling and accomplished little except to raise a cloud of dust which settled everywhere. Then I remembered that I was in a country where the expert use of dynamite was quite common knowledge. They used to say a good dynamiter in the Puget Sound country could blast a tree stump onto a dining room table if you asked him. We had such a man in the Public Works Department, and a few sticks of the right amount of dynamite correctly placed rendered the monolith susceptible to easy drilling.

On another occasion I introduced into the Navy Yard the new cutting alloy, carboloy. We had the job on the battleship *California* of cutting the top half of the conning tower off and installing an armored extension on the remains of the conning tower. The job called for a machine fit, so cutting with a flame was impossible and cutting case-hardened armor with ordinary steel cutting tools would take forever. I had heard of the introduction of carboloy in industry, so I bought enough to do the job at a fabulous sum per pound. The result was speedy work and a fine fit. The high-priced carboloy paid off.

I was able to introduce spray painting at the Puget Sound Navy Yard without any trouble with the union, and it certainly makes a difference in the appearance of a new ship. So far as I know, the *Louisville* was the first spray paint job that ever went to sea. I insisted that all my painters be tested for lead poisoning once a week.

In the summer of 1931 I went to the Bureau of Engineering and began putting some of my Navy Yard experience to work in the Bureau. When I reached the Bureau of Engineering I realized that, except in design, I had gotten about all the education in engineering I was going to get and that, more and more as time went on, I would have to play the game with what I had. I was in the Bureau for over eight years as Assistant Chief and Chief.

One of the first things that attracted my attention was contract dates of completion for material to be furnished and the penalty clause for failure to deliver on time. After much digging around, I found that the dates of completion set by the Bureau were pulled out of a hat and bore no relation to fact or to urgency. Further inquiry among the bidders' agents revealed just what I had suspected. The agent got a date of completion from his principal which represented the actual time to produce and, if that date followed the contract completion date, the bidder figured his penalty and included it in his price bid with the result that the Government, when it took a penalty, simply got its own money back. From then on I insisted on realistic dates of completion.

I set up an efficient and dominating Budget Office, one which was responsible for strict accountability and complete information in regard to funds.

Imitating to a certain extent the Schedule, Progress, and Accounting Section I had established in the Production Department at the Puget Sound Navy Yard, I set up in the Bureau a Scheduling, Estimating, and Progress Office. This Office was especially useful in working out building schedules for the construction of new ships. These schedules showed when all the machinery components of a ship should arrive at the ways or at the dockside to facilitate the rapid construction of a vessel. This Office also estimated the costs of new construction so that, when bids were opened, I was able to tell the Secretary of the Navy whether the bid was too high, reasonable, or too low. A very low bid usually represents an incorrect estimate by the contractor unless he is determined to get the bid at any cost.

We estimated the costs of tankers so successfully that Federal Ship sent their Chief Estimator down to the Bureau to see how it was done. In general, shipyards were far from brilliant in estimating or cost accounting. Finally I established an Office of Research and Investigation, under the late Captain Lybrand P. Smith, which concerned itself solely with accumulating information with respect to the latest trends and developments in technology, in industry, and in the scientific world. The scheduling and progress section and the research section were the first offices of their kind in the Navy Department.

While I was in the Bureau I had an opportunity to standardize illumination in our ships. I had long been interested in illumination, and in all our new construction I supplied the builders with an illumination chart showing the amount of illumination which must be supplied in every part of the ship, not forgetting the operating table in the sick bay.

I also originated the first acoustic survey made aboard ship in any Navy. In the building of the aircraft carrier *Essex* I was presented with a real problem. How best to transmit information to the pilots of a deckload of planes with engines running, just before a plane takes off? Many methods were considered—radio, telephones operated inductively from cable laid in the deck, and loud speakers. The Radio Corporation of America was hired to investigate this work and an acoustic survey indicated on what frequency the commands could be broadcast on a loud speaker to penetrate the bedlam and reach the ears of the pilots.

Because these acousticians were available, I used them to investigate the *Essex*, the noisiest ship I had ever been on. The noise emitted by the reduction gears was the worst I had ever heard, even worse than the gears of the World War I destroyers, which could be heard miles away. The noise from the *Essex* gears was 125 decibels,[5] the same as from

[5] The loudest industrial noise is 134 decibels measured 50 feet from the tail assembly of a jet engine. The human voice can be satisfactorily heard between 40 and 90 decibels. At 120 decibels one experiences annoyance and discomfort. At 140 decibels the sensation becomes definitely painful. There is general agreement that the maximum safe noise is 85 decibels.—*Science Newsletter*, April 5, 1952. A Science Service Publication, Washington, D.C.

a plane with an open cockpit. Conversation could not be maintained even by yelling in a man's ear. This was a totally unsatisfactory situation, and I required the shipbuilders who had cut and furnished these gears to replace them as defective. Such action was unheard of, and my popularity sank to a new low with the shipbuilders, if that was possible. It seemed as if the world never would learn that amateurs cannot hob reduction gears.

But the *Essex* had more noise trouble, a phantom noise which always seemed to be in the next adjacent compartment. It just couldn't be located. Dr. Harvey Hayes and his Sound Division of the Naval Research Laboratory solved that one. The frequency of the noise matched the normal period of the propeller. The energy was supplied when the blunt edge of the propeller hit the water. The sound was transmitted to the ship via the propeller strut and the propeller shaft.

When I became Chief of the Bureau of Engineering I transferred the Naval Research Laboratory, which had been under the control of the Radio Division of the Bureau of Engineering, to my own office and made it directly responsible to me. While I believe I was the first Chief who even attempted to follow the developments in radio, at the same time I used every effort to prevent the Research Laboratory from becoming just another communications laboratory. My pursuit of radio and radar was interrupted only by the enormous demands made on my time by the high-pressure high-temperature steam and top-heavy destroyer fights.

2

High-Pressure High-Temperature Steam

THE *Hammann,* Destroyer 411, epitomized the end of one era and the beginning of another—the end of the old-fashioned ship-and-engine builder and the beginning of the assembly type of shipyard.

In June 1939 the *Hammann,* equipped with steam machinery designed to employ pressures of 600 pounds per square inch and temperatures of 850°F, conducted trials at sea. On that bright June morning when the *Hammann* left the dock of the Federal Shipbuilding and Drydock Company, Kearny, New Jersey, where she was built, she had been subjected to only the usual dock trials while secured to a wharf. On that day, without any other preliminary trials, she proceeded to sea, ran a full-power run, went from full-speed ahead to full-speed astern and, at full-speed ahead, was repeatedly subjected to hard-over helm.

At the speed of 40 knots she stopped dead in 50 seconds, and 20 seconds later she was making 22 knots astern. She turned on the proverbial dime and was quiet and free from vibration. Moreover, she had space for two or three times the armament of previous destroyers. It is extremely doubtful that a new vessel of any kind had ever before been subjected to such rigid tests without preliminary trials at sea.

The performance of the *Hammann* illustrates the progress which the Navy had achieved in marine engineering up to that time and the progress made by the various manufacturing subcontractors with the component parts of ship machinery. Actually, this exhibition of the Navy's accomplishments in engineering was a publicity stunt arranged by

Assistant Secretary of the Navy Charles Edison with the help and encouragement of the Westinghouse Electric Corporation. When she shoved off from the dock on that memorable morning, the *Hammann* carried representatives from all the leading newspapers and magazines of New York City. The public demonstration of naval accomplishments was necessary in order to convince some of the rather severe critics in the press of the real progress that had been made. Needless to say, after such a remarkable demonstration of the success of naval engineering, the commentators were silenced and one heard no more about the subject.

How was the Navy able to achieve such extraordinary development in marine engineering, a development which has been copied far and wide, particularly by the U.S. Maritime Commission and the merchant service in general? It is the purpose of this chapter to trace this development and to describe how it was accomplished over unbelievable opposition and over the obstacles of blindness, stupidity, selfishness and even malevolence.

When I entered the Naval Academy in 1901, the usual prime mover was the reciprocating steam engine without forced lubrication.

In 1884 Charles A. Parsons (later Sir Charles) had introduced the steam turbine. It was a noncondensing reaction turbine, operating at 60 p.s.i. and developing 10 horsepower. It was the greatest invention in steam engineering since the days of James Watt. In 1891 Parsons added a condenser. In 1896 and 1897 Charles G. Curtis, an American, patented an impulse turbine which he admitted was a development of the work of Parsons and deLaval. Parsons first applied the turbine to marine use in 1897 on the *Turbinia*, a 100-ton ship of 2,100 horsepower. This demonstration of the possibilities of the steam turbine was the sensation of the day. The first turbine-driven ship in the Royal Navy was the cruiser *Amethyst* in 1904, equipped with Parsons turbines. Similar equipment was featured in the transatlantic liners *Lusitania* and *Mauretania* in 1907.

In the U.S. Navy it was decided to try the turbine in

competition with the reciprocating engine and, accordingly, three unarmored scout cruisers were built with the following characteristics:

Birmingham—two vertical inverted reciprocating steam engines, triple expansion
Chester—four Parsons turbines
Salem—two Curtis turbines.

These cruisers were all commissioned in 1908. The comparative tests were not considered conclusive. Many destroyers were built around 1910-1912 with turbine installations, but it was apparently considered inadvisable to install them in a ship of the line, and as late as 1911 the battleship *Oklahoma*, an oil burner, was authorized to be built with reciprocating engines. She was the last steam-reciprocating combatant ship for our Navy.

Meanwhile, in 1910, Charles A. Parsons made another contribution to the development of the turbine and its application to marine propulsion. A ship's propeller is not as efficient at high speeds as at low speeds, while the turbine is more efficient at high speeds. In 1910, Parsons performed a large-scale experiment on the steamship *Vespasian* which was equipped with a reduction gear between the turbine and the propeller. All previous turbine installations had been directly connected via the line and propeller shafting to the propeller. This reduction gear, both wheel and pinion, was cut with double and opposing (i.e. right-hand and left-hand) helical threads in order not to introduce thrust. This innovation was a marked improvement, and reduction gears are used to this day for marine and certain land applications.

Obviously the steam turbine was still in the developmental stage and had a long way to go. The theory of torsional vibrations was not enunciated until 1919, and for that reason we would continue to break propeller shafts for many more years. The theory of nodal vibrations of turbine blades was unknown until 1925, and no easy means existed to attain dynamic balance. That lack of knowledge did not make too much difference, however, because there wasn't very much right about the first turbines, either Parsons or Curtis. Radically new and progressive inventions are always discouraging at first. The first model always ends up in a museum.

In my early days, coal burning was universal, although a great deal of attention was being given to the development of the art of fuel-oil burning. No one knew anything about the treatment of feed water for boilers, and very little was known about the design of boilers, either the old Scotch cylindrical type or the brand new water-tube type. Even as late as 1931 the possibilities of natural or gravity circulation in water-tube boilers were not dreamed of. The use of forced draft was usually accompanied by frequent breakdown of the forced-draft blower engines. Refrigeration, if any, depended on the old-fashioned Allen dense-air ice machine.

We were still calling electric generators dynamos. Our ships were wired for direct current. As a matter of fact, alternating current was not introduced aboard any naval vessel until 1934, and the U.S. Navy was the first to take alternating current to sea for auxiliary use. The main drive of the turbo-electric ships was, of course, alternating current.

The airplane was not much considered; the Navy's Bureau of Aeronautics was not established until 1925. In 1926 General Billy Mitchell was court-martialled for his advocacy of air power, and in 1938 I thought my own career might end the same way for my support of high-pressure high-temperature steam.

Modern accounting, the principal tool of the industrial manager, was not introduced in the Navy until about 1910. Frederick Taylor of time-study fame was creating a sensation with his book, *The Art of Cutting Metals.* Gantt and the efficiency experts were rampant.

In the early 1900's the gas engine, which was to become predominant in our small boats and submarines until superseded by the diesel engine, had not yet appeared in large numbers, and when it did appear it was so unsatisfactory that no one wanted it.

Not so long ago radio was still in the spark-gap stage and only short-range communication was possible. We were still trying to eliminate static by hitting the coherer with a hammer.

Our own country was not emancipated from Europe, especially England, commercially, industrially, or in engineering.

When we compare the equipment we had to work with fifty years ago with the equipment we have at our disposal today, we should remember that the unbelievable strides we have made in the mechanical and electrical arts have been due to the amazing contributions of individuals who were scientists, research workers, or development engineers. None of these things just happened. These improvements were the result of long, hard, and generally disheartening work by many unsung heroes in research and development. For instance, General Electric acquired the rights to manufacture the Curtis turbine when the Curtis turbine was a whole lot more of an idea than it was a turbine. It took a great many years and millions of dollars' worth of effort to arrive at the dependable steam turbine of today.

The expression "high-pressure high-temperature" covers all the improvements made in the propulsion machinery in our new naval designs from 1933 to 1939. To appreciate thoroughly the value of these changes, to understand why the U.S. Navy and not the Royal Navy pioneered in this great advance in naval engineering, and to understand the blind and violent opposition to these changes, one must consider and analyze a considerable background of circumstances and fact.

Britain had been the leader in naval affairs for a century. This leadership stemmed from her leadership in marine engineering and naval architecture and from her leadership in industry. *It is an axiom that no navy can rise superior to the technological ceiling of its own national industry.* The date of our political emancipation from England is well known and justly celebrated. On the other hand, we attained our engineering and industrial independence so slowly that the phenomenon scarcely attracted attention. By the time the rebuilding of the U.S. Navy began in 1933, we had not only completed our industrial emancipation but had taken the lead in many engineering and industrial lines.

It had been recognized for generations that we could reduce the size of our machinery and increase its efficiency if we could increase the pressure and the temperature of the steam. Such increases had to await advances in metallurgy in order to withstand high-pressure high-temperature conditions.

The greater the heat-drop in any machine, the greater the efficiency. To utilize this heat-drop, steam must expand from small to large volumes. This requires that, if we desire to reduce the size of machines, we must start with high pressure. If we start with a high pressure and do not have an accompanying high temperature, the expanding steam soon reaches a moisture content impractical for use in turbines.

One method of avoiding such moisture content is reheating. That is, when the steam has expanded to a certain extent it is taken out of the turbine, passed through a heater, and finishes its expansion in another turbine. By this means we can theoretically increase the efficiency almost indefinitely without at any time exceeding any given temperature at the high-pressure end or any given moisture content at the low-pressure end. The great complication of piping and apparatus required for reheating, however, precludes its widespread use in naval vessels, and its extensive use in fixed central power plants is quite recent. The only alternative is to start with a temperature high enough to avoid excessive moisture in the exhaust.

The basic stimulus to the utilization of the widest practicable temperature range came with the publication of Sadi Carnot's classic paper in 1824. The efficiency of the Carnot cycle is given by:

$$E = (T - T_0)/T = 1 - T_0/T$$

Where E = efficiency, T = highest, and T_0 = lowest *absolute* temperature available.

The Carnot cycle as thus expressed is based on the assumption of an ideal gas. Now steam is not an ideal gas; when cooled, it does not even remain a gas. In practical use it goes from gas to liquid and back to gas again with all the complications incident to change of state. The heat that steam may possess in excess of the amount necessary to maintain its existence as a dry gas at a given temperature is superheat.

Rankine, in 1885, gave a definition of efficiency which would apply either to an ideal gas or to a real substance such as steam; it also applies whether or not the theoretical maximum efficiency is being obtained.

$$\text{Thermal efficiency of transformation of heat into work} = \frac{\text{heat added} - \text{heat rejected}}{\text{heat added}}$$

It is obvious that, from either of these equations, higher initial operating temperatures mean higher efficiency.

How high in temperature and pressure we should go is another question. The higher we go the greater are the possible efficiencies, but there is a law of diminishing returns. The first increases of pressure and temperature produce great gains in efficiency. Further corresponding increases produce gains that are not so great. Study of efficiency curves for various temperatures and pressures shows that gains are quite rapid up to between 800° and 900°F and between 600 and 800 pounds per square inch pressure, and thereafter gains continue more slowly.[1]

The initial temperature of the steam is primarily limited by the state of the metallurgical art. Since weight of machinery usually increases with pressure, we have for shipboard installations a limiting factor on pressure. The final temperature is determined by the temperature of the sea water used as a coolant in the condenser. This may be 80°F or slightly more in some oceans and a little lower than 32°F in others. There are other practical considerations such as

[1] See Appendices A and B.

the size of the condenser, which limits the volume of exhaust which can be handled irrespective of the terminal temperature. Obviously, the best way to raise the efficiency of a heat engine of the turbine type is to raise the pressure and temperature of the entering steam. Let us examine the situation of the shipbuilding and engineering world in 1933 at the beginning of the Navy's building program.

Ships are built in navy yards or in private shipyards under contracts approved and signed by the Secretary of the Navy. The General Board of the Navy, an advisory group responsible to the Secretary of the Navy, establishes the military characteristics of all new ships. Military characteristics include principally the type, length, breadth, depth, tonnage, compartmentation, speed, cruising radius, armor, and armament. In general, the Bureau of Construction and Repair submitted to the General Board for approval the preliminary plans for the vessel's hull and appurtenances, the Bureau of Ordnance likewise submitted plans for armor and armament, and the Bureau of Engineering for propulsive and auxiliary machinery such as electric generators, refrigerating machinery, and radio equipment.

The Chief of every bureau is appointed by the President with the advice and consent of the Senate. This is also true of the Chief of Naval Operations. These are all statutory positions in contradistinction to membership on the General Board, where vacancies are filled by the Bureau of Naval Personnel. After a ship is built, various official trials are held before the ship is accepted. These trials are conducted by the Board of Inspection and Survey whose membership is supplied by the Bureau of Naval Personnel, as in the case of the General Board. It is the responsibility of the Board of Inspection and Survey to see that the terms and specifications of the contract and the guarantees are met with respect to new vessels; it also recommends alterations and changes which in its opinion are necessary to make the new vessel an efficient unit of the fleet.

Before bids for the construction of a Navy ship were requested, the Bureau of Engineering would prepare special specifications for the machinery installation. In these special specifications the characteristics of the ship as to power and speed were laid down together with the steam conditions to be used and the general design of the main propelling plant. In addition, contract plans were prepared showing the general arrangement of the machinery in the spaces provided, major piping systems, etc. These special specifications and contract plans, together with the General Specifications for Machinery, formed a part of the contract for the construction of the ship. No departures from these basic design features were permitted without the specific approval of the Bureau of Engineering and, in major cases, the Secretary of the Navy.

The shipbuilders based their bids on these special specifications, contract plans, and general specifications. The Bureau of Engineering did not furnish the details of the design of the various machinery units or the working plans necessary for their fabrication. It was necessary, therefore, that the shipbuilder either prepare or have prepared for him the details of the design of the machinery units and the working plans necessary for their fabrication. This he must agree to do when submitting his bid.

From 1918 to 1933 the progress made in naval engineering was not great. A relatively small number of ships were built during this period and they showed no marked advance in their design. All of the cruisers built during this period, with the exception of four designed and constructed by the Navy, used saturated steam. One class of destroyers, the *Farragut* class, used steam at 650°F. The designs of the main propulsion turbines, reduction gears, boilers, feed systems, etc., were practically unchanged from those used during the World War I building program.

All of the Navy ships built under private contract during this period (1918-1933) were designed by the "Big Three" —Bethlehem Shipbuilding Corporation (Bethlehem Steel), Quincy, Massachusetts (formerly Fore River); Newport News Shipbuilding and Drydock Company, Newport News,

Virginia; and New York Shipbuilding Corporation, Camden, New Jersey. The design of these vessels, while controlled by the specifications issued by the Bureau of Engineering, reflected to a large degree the design policies of these companies, which had been very conservative in their designs, particularly New York Ship, and had made little advance during the period under discussion. This lack of progress on the part of these shipbuilding companies was probably due to three fundamental causes: first, the small number of ships being built; second, the lack of research facilities in the shipbuilding industry; and third, dependence on Parsons, Ltd., for engineering. The holiday in naval, as well as merchant, ship construction was not conducive to any great improvements in design. The shipbuilding industry is always in a feast or famine condition, and the country was still in the depression—including the shipbuilding industry. The industry did not appreciate the fact that the old ship-and-engine builder was gone and that shipyards must become assembly yards. They could no longer compete with the highly specialized manufacturers of turbines, reduction gears, boilers, etc., who spent millions for research and experiment before they marketed their products.

All of the "Big Three" held licenses from Parsons, Ltd., for the fabrication of turbines. Parsons carried out what research was done for these shipbuilders, reviewed their turbine designs and furnished them with new developments, if any. This policy resulted in the British Admiralty having made available to it the power, speed, and the designs of turbines being installed in American naval vessels. The shipyards were entirely dependent upon Parsons, Ltd., for any research in engineering, which at this time in turn reflected the inbred conservatism of British engineering. The result of this was that the main turbines fabricated by the "Big Three" were substantially the same year after year and showed practically no advance over the turbines built by these companies for World War I construction. In 1935 the Chief of the Bureau of Engineering, Rear Admiral (later Admiral) S. M. Robinson, notified the "Big Three" shipbuilders—with my concurrence as I was to relieve him as Chief a few days later—that the Bureau had decided to enforce the provisions of the Espionage Act and that therefore they would have to sever their licenses under Parsons, Ltd., if they wished to proceed under any future Navy contracts. This stopped the siphoning of U.S. Naval turbine designs to England and made us dependent on American industry for our machinery designs. This was a vital step in the development of American naval power.

Prior to the creation of the first steam power plant or central station by Thomas A. Edison in 1882 and for some time afterward, the navies of the world were the leaders in the development of steam engineering. Finally the central power plants became the exponents of the most advanced practices in the art. The central power plants were usually steam plants located in or near large cities, although there were many hydro-electric plants, often with long transmission lines. The steam plants generated huge quantities of electricity for the light and power requirements of their areas.

During the period following World War I, the design of turbines and boilers used in central stations in this country had progressed very rapidly, and this advance had been accompanied by a marked increase in the efficiency of these plants. Higher and higher steam pressures and temperatures had been employed until 600 pounds, 850°F steam was commonplace. In many cases much higher pressures and temperatures were often employed. The turbines and other machinery units used in central power stations had been developed by private industry but outside of shipbuilding. General Electric, Westinghouse, Allis-Chalmers, and deLaval were among the leaders. These companies, particularly General Electric and Westinghouse, had very large research and development facilities and had spent large sums on the development of turbine designs. As a result, this country had in operation in our central power stations the most modern and efficient turbine installations in the world. These turbines represented a distinctive American contribution.

The development of boilers in shore power stations had also been very rapid. Here again private industry, but not shipbuilders, was responsible and had made possible the advances by large expenditures for research. There had not been any comparable development in marine boilers.

The main propulsion turbines designed by the "Big Three" shipbuilders under their Parsons licenses were characterized by their relatively low speed, large size, and large number of blades. They were typical of the designs then used in England in both naval and merchant ships. Such turbines were not suitable for use with high-temperature steam due to difficulties caused by distortion of casings and rotors, with resulting damage. The turbines developed by the General Electric and Westinghouse companies for central stations were characterized by their high speed, small size, and small number of parts. Such designs were adaptable to the highest steam temperatures contemplated, were very rugged, and free from distortion.

In 1927, when the contracts for armored cruisers CA 26-31 were awarded, the Bureau of Engineering attempted to raise the steam conditions by specifying 300 pounds per square inch gauge and 100°F of superheat (about 520°F) as an alternative specification. Bids were received responsive to that specification in addition to the basic bids on 300 pounds saturated steam (about 420°F). A conference of successful bidders was held in the Bureau of Engineering. They recommended that no superheat be used, and this recommendation was approved. As a result, the machinery of these ships was out of date before they were built. It was not until the *Farragut* class of destroyers was contracted for in 1931 that any real advance in steam conditions was obtained. Here, 400 pound, 650°F (about 200°F of superheat) steam was specified by the Bureau and built into the ships. This design was prepared by Bethlehem. The *Porter* class of destroyers was contracted for the following year, and again 400 pound, 650°F steam was specified and obtained. This design was prepared by New York Ship. In both the *Farragut* and *Porter* classes of destroyers the large, slow-speed type of turbines, characteristic of the designs of these companies and of the British Navy, were installed together with single reduction gears and the conventional integral superheater type of boilers without economizers.

In 1933 President Roosevelt, alarmed at the international outlook, took $281,000,000 of National Industrial Recovery Act (NIRA)[2] money and started rebuilding the Navy. This was one of the finest and most statesmanlike things Roosevelt ever did. He knew exactly what he was doing because, afterwards, he said to his Secretary of the Navy, Claude A. Swanson: "Claude, we got away with murder that time." Perhaps it should be added in explanation that, at the time, pacifism was so strong in our country it was only possible to begin rebuilding the Navy by using underhand methods. This move of Roosevelt's put a lot of men to work, so Congress and the public swallowed it.

At about the time that Roosevelt made his speech in Chicago in October 1937, about building a "cordon sanitaire" around those nations which were provoking a war and which produced an unfavorable reaction in Congress, I had a significant talk with his personal secretary, Marvin McIntyre. I asked Mr. McIntyre why the President didn't take credit for his rebuilding of the Navy with NIRA money. McIntyre told me to forget it because the pacifists were too strong to make it safe to let the public know anything about it.

The Mahan Class of Destroyers

As a result of this Naval Building Program a large number of projected destroyers would have been built very similar to the *Farragut* class, contracted for in 1931, if there had not

[2] The National Industrial Recovery Act authorized the Administrator of Federal Emergency Public Works, under the direction of the President, to undertake specific programs, one of which was Naval shipbuilding. The Fourth Deficiency Act, Fiscal Year 1933, appropriated $3,300,000,000 for the purpose of carrying out the NIRA authorization act. The Administrator of Federal Emergency Public Works, established under Title II of the NIRA Act upon approval by the President, granted a request received from the Secretary of the Navy for the allocation of shipbuilding funds. Throughout the life of this allocation, the Navy reported

occurred a sharp break with precedent. As a result of the bidding for these destroyers, contracts were awarded to Bath Iron Works, United Shipyards, and Federal Shipbuilding and Drydock Company (U.S. Steel) for two ships each. If the "Big Three" had been the successful bidders, the destroyers never would have been built to specifications. The ships were to be identical in their design. It was necessary, therefore, that one of these three companies develop the design and prepare the working plans for the entire class. None of these bidders had what the Bureau of Engineering considered to be an adequate design and drafting organization. They were directed, therefore, by the Chiefs of the Bureau of Engineering and the Bureau of Construction and Repair to establish such an organization or to hire one. They decided to employ the firm of Gibbs and Cox for this work.

The firm of Gibbs and Cox had had no experience in the design of Naval vessels. They had had, however, wide experience in merchant ship building for the preceding fifteen years. They had designed the *Santa Rosa, Santa Paula, Santa Lucia* and *Santa Elena* for the Grace Line. These ships were built by Federal Shipbuilding and Drydock Company of Kearny, New Jersey, and they were designed for a steam condition at the turbine throttle of 375 pounds gauge, 300° superheat, or a total temperature of 743°F. The high-pressure turbine operated at 4,500 rpm and the low-pressure turbine at 3,500 rpm. The boilers were equipped with feed water heaters or economizers and superheaters of the integral type. The turbines were connected to the propellers through double reduction gears. Thus it is seen that these vessels embodied engineering practices far in advance of anything that had been attempted in the U.S. Navy. Therefore the Bureau of Engineering was particularly anxious to obtain the benefit of this organization's experience in modern marine engineering.

---

expenditures and obligations regularly to the Administrator of Federal Emergency Public Works. The opinion of some that this money came from the Public Works Administration (PWA) is not correct.

When the successful contractors agreed to employ Gibbs and Cox as their design agent, it was decided to install in these 1933 destroyers the most modern types of machinery available. It was realized at the time that this was a radical step to take and that mistakes were bound to occur, but it was essential to take this step if the new Navy, then being built, was not to have machinery installations which would be obsolete before the ships were launched. These decisions resulted in the *Mahan* class of destroyers, the first type of vessel incorporating any of the modern engineering designs in our Navy. Concurrently with the decision to install the most modern types of machinery available, and in collaboration with the General Board of the Navy and the other Bureaus, a weight and space limitation was accepted for the machinery installation which resulted in too much congestion. This became apparent before the ships were completed and, as a result, more space was provided in succeeding ships. This congestion was in no way connected with the modern type of machinery used but was due solely to attempting to install more horsepower in a given space than had ever been attempted before. The Navy would still have to learn that progress and displacement are closely related, that more progress usually means more displacement, and that the sardines are more important than the can.

Table I shows a comparison of the design features of the *Farragut* class (Bethlehem design) and the *Mahan* class (Gibbs and Cox design). The *Mahan* class, it will be remembered, were the destroyers contracted for in 1933 while the *Farragut* class were contracted for in 1931.

The machinery installation of the *Mahan* class marked the first time that high-speed turbines with double reduction gears and direct coupled cruising turbines had been installed in a Naval vessel. These turbine speeds are much greater than power plant turbine speeds because 60-cycle power plant turbines are restricted, necessarily, to a speed of 3,600 rpm. This main machinery was designed by the General Electric Company and into its design went all of the experience of that company with land as well as marine turbines, together

TABLE I

	Farragut	*Mahan*
Steam conditions	400 pounds 650 degrees F	400 pounds 700 degrees F
Boilers	three drums integral superheaters no economizers no air casings	three drums integral superheaters economizers no air casings
Turbines	2-turbine arrangement Speeds: high pressure 3,460 rpm low pressure 2,320 rpm	3-turbine arrangement with direct coupled cruising turbines (no clutches) Speeds: high pressure 5,850 rpm low pressure 4,926 rpm cruising 10,000 rpm
Reduction gears	single 2 pinions	double 2 pinions
Feed Water System	non-deaerating	deaerating

with the developments brought forth by their research facilities. This machinery was the most rugged and reliable of any main drive installation ever installed in the Navy up to that time. These ships represented a complete American design in all respects.

It was unfortunate that so many ships (26) were involved in this one design, the *Mahan* class. The conditions under which this class was laid down were never duplicated before in naval history. The building program was started in the midst of an unprecedented depression with the avowed purpose of furnishing employment. The dominating and only idea at the time was to put men to work as rapidly as possible, and almost every week there was a conference in the office of the Assistant Secretary of the Navy, Colonel Henry Latrobe Roosevelt, which was attended by the responsible Navy officers, representatives of labor, and representatives of the administration in which the material bureaus (Ordnance, Engineering, and Construction and Repair) were constantly criticized for the delays in reemployment on account of delays in plans, delays in material, etc. Because of the long building holiday, the material bureaus, as well as the shipbuilders, were handicapped by the scarcity of draftsmen, there being at that time one-third the number which were available in 1938. Those who are unfamiliar with this situation or who did not participate in it first-hand cannot possibly realize the amount of pressure applied to the material bureaus to get the building program underway and provide employment with the greatest possible speed and at whatever cost.

The *Mahan* class proved exceptionally reliable. Cost of upkeep was normal, and the machinery was easy to operate. It was originally planned for 400 p.s.i. and 850°F, but fear of the autogenous[3] temperature of ordinary lubricating oil, and perhaps a desire to appease the critics of high temperature who had already appeared, resulted in completion of this class for operation at only 700°F.

When I relieved Rear Admiral S. M. Robinson on 29 May 1935 as Chief of the Bureau of Engineering and Engineer-in-Chief of the Navy, the Bureau was in a rather tranquil mood. Robinson had started many things but they had not proceeded far enough to excite any opposition. Some of these innovations will be covered in other chapters. Not long after I became Chief there was a rising tide of criticism of the *Mahan* class of destroyers because of crowded conditions in the machinery spaces. There was also a great deal of criticism about the number of so-called gadgets which were being introduced aboard ship, and a lot of time was consumed uselessly in criticism of weights of machinery and

[3] The autogenous temperature of lubricating oil is about 850°F. It varies with pressure, container, concentration of vapor, as well as with the kind of oil used.

electrical installations, particularly the rugged type of junction box used in naval wiring. But the criticism of the *Mahan*-class gadgets and electrical clocks aboard ship was nothing compared to the storm which broke following the preliminary acceptance trials of the destroyer *Somers*, DD 381, in October and November of 1937.

The Destroyer Somers

It is now necessary to consider some of the improvements which were embodied in the *Somers* in order to evaluate the storm of criticism which these improvements evoked.

Boilers. The great improvement in the boilers of the *Somers* resulted from several causes. Some time previously, after re-engining the *New Mexico*, it was found that the new boilers delivered more superheat than had been planned for when the new turbines were designed. The boilers were furnished by one firm and the turbines with attendant condensing apparatus by another. Serious distortion and other damage, sustained shortly after the sea operations of the *New Mexico*, required alterations of the low-pressure end of the turbine and the condensing apparatus. The contractor claimed that he should be reimbursed for these unforeseen costs since the initial and therefore final temperatures were considerably higher than the temperatures for which he had contracted to build the turbines and associated apparatus.

The boilers of the *New Mexico* had what is known as an integral superheater; that is, all the steam generated passed through a superheater heated by the same gases of combustion. Not only did the amount of superheat vary directly with the output of the boiler, but there was no way to tell how great this superheat would be. This was amazing. In other words, the amount of superheat could not be controlled, nor was there on the market at that time any suitable method of accurately controlling superheat. Accurate control is necessary for not only design reasons, as shown in the case of the *New Mexico*, but also because marine turbines are subject to many and various changes in speed incidental to maneuvering while a ship is underway. High superheat, necessary for economy when a vessel is in free route and making good speed, may become a handicap when the main engines are slowed down or backing.

Power plant engineers ashore already were controlling the amount of superheat in their boilers by the use of dampers fitted in the uptakes. Dampers cannot be used in naval boilers because naval boilers are forced to a greater extent than power plant boilers, and dampers would not stand up under such forcing.

Babcock and Wilcox and Foster-Wheeler responded to my demand for superheat control by designing a superheat control boiler which permitted such rapid variations in degrees of superheat that the changes had to be recorded by a thermocouple rather than a mercury thermometer.

This boiler was "air-encased"; that is, it was surrounded by two casings with air in between—the incoming air on its way to furnish combustion in the fire box. With this boiler it was possible to operate with an "open" fire room (with the hatches open). In a "closed" fire room the air for combustion is discharged into the fire room as a whole and passes through the fuel oil burner doors into the fire box.

In case of gas attack with a closed fire room, the boiler room attendants are helpless since they breathe the same air that is used for combustion. With an open fire room the gas-laden air is all fed to the boilers, and the air in the fire room can be kept free from contamination. The open fire room is also cooler and quieter.

It should be noted that these *Somers* boilers had been operated at 40 per cent overload at the Boiler Laboratory in the Philadelphia Navy Yard before installation with no signs of breakdown of natural (i.e. gravity) circulation. Consequently, in an emergency, three boilers overloaded could do the work of four boilers operated normally. This in itself was a new safety factor to lessen the effect of battle damage.

I will later describe the reception this boiler was accorded when it appeared in the *Somers*, the first vessel to operate at 600 p.s.i. and 850°F in the U.S. Navy.

Turbine Testing. Just as the *Somers* boiler was given elaborate tests ashore in the Naval Boiler Laboratory prior to installation, so was the entire propulsion machinery for one shaft of the destroyer *Benham,* whose machinery was quite similar to that of the *Somers.* But the Bureau of Engineering did not have a turbine-testing plant large enough to handle main turbines, so these turbines were tested under steam at full power and various loads ahead and astern for long periods at the South Philadelphia plant of the Westinghouse Corporation where the main turbines of the *Benham* were built. In fact, in this shop these main turbines operated at full power for longer periods than might normally be required for several years in times of peace. At cruising and other speeds the operation represented many, many round trips to Europe. I doubt if any such full-scale shop tests on main propelling machinery were ever carried out anywhere in the world before.

Curtis Turbines. A great deal of research and development work on turbines had been done since World War I. The new high-speed General Electric Curtis turbine carried 1,750 blades as against 17,500 blades in the slow-speed Parsons turbines furnished by the "Big Three." The rotor was usually 25 per cent shorter in the high-speed turbine and was machined from a forging while its competitor was a built-up affair. Anyone with a good engineering eye could recognize in an instant the superiority of the forged rotor.

In 1941 the patent of Charles G. Curtis for his impulse turbine was sustained in the U.S. Circuit Court of Appeals for the Third Circuit. Incidentally, the opinion of the court is a remarkably clear exposition of the difference between impulse turbines and reaction turbines.[4]

The Curtis turbine had a higher efficiency than the Parsons type and a new method of controlling speed. Land rights were acquired by the General Electric Company. The *Encyclo-*

[4] The U.S. Circuit Court of Appeals for the Third Circuit, #1622. *International Curtis Marine Turbine Company and Curtis Marine Turbine Company of the United States vs. William Cramp and Sons Ship and Engine Building Company.* Federal Report, Vol. 211, p. 124.

paedia Britannica[5] states that the company spent $3,000,000 in developing this turbine with the result that it became used widely ashore and afloat and in both the United States and Royal British Navies.

This $3,000,000 must have been a drop in the bucket compared with what was spent later and until about 1931 by the General Electric Company. The brilliant work of Mr. Wilfred Campbell, assisted by Mr. E. L. Robinson, revealed the fundamental cause of failure of turbine blading and developed an analytical method to guard against future failures. Campbell's work is covered in a classic paper presented to the American Society of Mechanical Engineers and entitled "The Protection of Steam Turbine Disc Wheels from Axial Vibration." Following Mr. Campbell's death in 1924, a second paper was prepared by Mr. W. C. Heckman from Campbell's notes and presented in 1925. This paper, entitled "Tangential Vibration of Steam Turbine Buckets," showed how the principles applying to disc wheel vibration applied also to tangential vibration of steam turbine buckets. Heckman was not content to stop there. Together with Robinson he developed a method of proportioning turbine buckets so as to enable them to carry the required load irrespective of the type of vibration present.

The practical results of this work were tremendous. In the early twenties, for example, every time the turbo-electric driven *New Mexico, Saratoga*, and *Lexington* came to the Navy Yard for overhaul it was necessary to renew the blading in the eighth stage of the rotor of the *New Mexico* and the fourteenth stage of the others. These technical advances solved that problem and eliminated for all time similar failures aboard ship and in central power plants.

Inspection of Castings. Great progress had also been made in the inspection of the interior of castings and hollow forgings. By 1928 Dr. R. F. Mehl of the Naval Research Laboratory had developed gamma-ray photography to the point where it was possible to discover in the interiors of castings

[5] Thirteenth Edition (1926), Vol. 29, p. 778.

and forgings defects which could not be revealed in the process of machining. X-ray photography was of course at that time in its infancy. Gamma-ray examination of castings was first made part of a Navy contract with the replacement of turbine castings for some of the *Farragut* class.

Carbon Molybdenum Steel. Metallurgists already had determined how much temperature and pressure carbon steel would safely stand and under what conditions carbon molybdenum steel should be employed. There was also plenty of carbon steel piping in oil refineries subject to very high pressures and temperatures, and which had stood the gaff for many years although later practice would have demanded carbon molybdenum. In new installations, beginning with the *Mahan*, carbon molybdenum steel was used where good engineering practice demanded it. Pipe flanges were welded together instead of bolted, and sweeping bends supplanted the old-fashioned slip joints to accommodate expansion and contraction of the piping.

Mock-ups and Modeling. In all this rapid progress in American engineering from 1918 to 1940, perhaps nothing is so characteristic of that progress and the resulting complexity of machinery and equipment as the rise and perfection of the art of modeling.

Because of the extremely crowded conditions in the engine rooms of submarines, it had long been a practice to make a "mock-up," as it was called, of the machinery in the engineering spaces. It would have been almost impossible to install all the machinery required for operation if this had not been done, and it would have been quite impossible to install it in such a fashion that it would have been easily operable or easily accessible for repair. This building of models for machinery spaces was also developed at the Brooklyn Navy Yard and was employed there in connection with the design and building of the battleships *North Carolina* and *Washington*, the first battleships of the new building program. Nowhere, however, did the art of model building exceed in excellence its development at Gibbs and Cox, the designers of so many of our controversial destroyers.

Models for the machinery and other spaces of destroyers, and later for all vessels for which they were the design agent, were constructed accurately to scale. No piece of equipment or member of the hull of the model was installed until the drawings were approved by the appropriate Bureau of the Navy Department. The model thus became, in this respect, a progress chart for plan approval. The models were so accurate that it was safe for a draftsman to obtain dimensions from them, and it was much easier for the pipe shop to lay out its piping from the model than to construct it by laying templates in the actual vessel. The improved arrangement and accessibility of destroyers after the *Mahan* may be credited in great part to model making.

The Navy Department was so impressed with this approach that it finally stipulated in its contracts that the shipbuilder must supply a model in addition to the ship he was building. It was hoped that this would result in the widest possible use of models with all the attendant improvement in planning design. New York Ship's Mr. J. F. Metten fought this innovation, almost to the point of losing a contract, although $50,000 was included in the contract for the cost of the model.

It thus appears that the Parsons' license was more than a license—it was a state of mind.

Preliminary Acceptance Trials of the Destroyer Somers

The destroyer *Somers*, DD 381, ran her preliminary acceptance trials at the Rockland, Maine, Standardization Course and, at sea, off Rockland, on 26 October to 4 November 1937.

The advance report of these preliminary trials by the Board of Inspection and Survey signalizes the real opening of the high-pressure high-temperature steam fight. The *Somers* operated on these trials at 600 p.s.i. and 730°F.

Let us compare the performance of the *Somers* and the *Porter*, based on the figures contained in the official trial board reports of those vessels. The *Porter*, DD 356, was contracted for in 1933 and was designed and built by the

New York Shipbuilding Corporation. The *Somers*, DD 381, contracted for in 1934, was designed by Gibbs and Cox and built by Federal Shipbuilding and Drydock Company (U.S. Steel). The *Porter* represented the latest type of destroyer to be designed and built by New York Shipbuilding Corporation up to that time. In her were employed low-speed turbines of typical British design, single reduction gears, and no superheat control. The *Somers* was representative of the modern American design employing high-speed turbines, double reduction gears and superheat control.

TABLE II

	Porter DD 356	*Somers* DD 381	Saving in fuel by *Somers* (percentage)
Trial displacement	2,157	2,143	
Full power trial, SHP (shaft horsepower)	47,271	51,525	
Full power speed, knots	37.17	38.85	
Steam conditions	400#, 700°F	600#, 700°F	
Lbs. fuel oil per sea mile:			
at 10 knots	142.5	111.0	22.1
at 15 knots	167.5	131.5	21.5
at 20 knots	222.0	181.5	18.3
at 25 knots	340.0	289.4	14.9
at 30 knots	569.0	497.5	12.5
at 35 knots	800.0	733.0	8.4
at 37.17 knots	833.5		
at 38.99 knots		925.0	
Cruising radius at 15 knots (sea miles)	8,710	10,540	

Somers' cruising radius was 21 per cent greater than *Porter's*.

Somers' increased economy was due to higher pressure and high-speed machinery.

For the same performance, the *Somers'* machinery weighed 10 per cent less than the *Porter's*.

It would be possible to build another *Porter*, using *Somers*-

type machinery, having the same steam conditions, producing identical results and save about 130 tons in the total weight of the ship. This weight is the equivalent of three twin gun mounts and necessary ammunition. Using 850°F steam, 236.9 tons weight of machinery and fuel would be saved, an equivalent of six twin 5-inch enclosed gun mounts together with necessary ammunition.

Next let us compare the *Gridley*, DD 380, and the *Benham*, DD 397. The *Gridley* was contracted for in 1934 and was designed and built by Bethlehem. She represented the latest type of destroyer to be built and placed in service by that company. Like the *Porter*, the *Gridley* was representative of the less modern type of machinery installation, being equipped with low-speed turbines of typical British design, single gears, and no superheat control. The *Benham*, contracted for in 1935, was designed by Gibbs and Cox and built by Federal. Her machinery installation was representative of the more modern American design, consisting of high-speed turbines, double reduction gears, and superheat control type boilers. The main turbines and reduction gears of the *Benham* were designed by Westinghouse Electric and Manufacturing Corporation.

Benham's cruising radius was 22.8 per cent greater than *Gridley's*.

Benham's increased economy was due solely to high-speed machinery.

For the same performance, the *Benham's* machinery weighed about 12 per cent less than the *Gridley's*.

This comparison of the trial performance of the *Gridley* and *Benham* proved the superiority of the new high-speed machinery even when the steam pressures and temperatures were identical.

It would be possible to build another *Gridley* using *Benham*-type machinery, having the same steam conditions, producing identical trial results, and save about 85 tons in the total weight of the ship. This weight is the equivalent of four single gun mounts and necessary ammunition. Using 850°F steam, 176.9 tons weight of machinery and fuel

TABLE III

	Gridley DD 380	*Benham* DD 397	Saving in fuel *by Benham* (percentage)
Trial displacement	1,771	1,762	
Full power trial, SHP	47,285	50,201	
Full power speed, knots	38.99	40.90	
Specified steam conditions	600#, 700°F	600#, 700°F	
Lbs. fuel oil per sea mile:			
at 12 knots	142.0	110.0	22.6
at 15 knots	152.0	125.0	17.75
at 20 knots	207.5	167.0	19.55
at 25 knots	338.0	298.0	11.85
at 30 knots	528.0	450.0	14.88
at 35 knots	700.0	585.0	16.4
at 38.5 knots (full power)	810.0	692.0	14.6
at 40.8 knots (full power)		775.0	
Cruising radius at 15 knots (sea miles)	7,735	9,500	

would be saved, an equivalent of eight 5-inch single enclosed gun mounts together with their allowance of ammunition.

While the foregoing tables are based on the same steam conditions, as a matter of fact both the *Somers* class and the *Benham* class were built for operation with 850°F steam in accordance with my orders. The *Somers* had operated under this condition at that time. Neither the *Porter* class nor the *Gridley* class could operate at this temperature. By taking advantage of the 850°F steam, the *Somers* and *Benham* developed about 10 per cent more power, more than a knot greater speed, further improved their economy about 6 per cent, correspondingly increased their cruising radius, and reduced the unit weight per horsepower by approximately 10 per cent.

Another comparison of interest will be discussed briefly

here; that is, the question of reliability of the two types of design. In this respect the advantage again lies with the more modern design. The turbines contain but a fraction of the number of blades used in the old design—14,730 in the *Somers* versus 98,750 in the *Porter*. The turbines are much shorter—from 25 to 33 per cent. This makes them less likely to distort with changes in temperature. The turbine rotors are solid forgings instead of the hollow built-up design used in the less modern type, again reducing the danger of distortion due to temperature changes. Higher factors of safety have been used throughout the modern turbines than in the older types.

The double reduction gears have higher factors of safety than have the single reduction gears used with the older design. In addition, they are much quieter.

The boilers used in all the classes of destroyers of the modern design represent three distinct steps in boiler design—superheat control, economizers, and air casings. All of these boilers were thoroughly tested at the Naval Boiler Laboratory prior to their installation in the ships. This was not done in the case of certain boilers used with the older designs, and operating difficulties developed in service which required corrective measures.

As a result of these trials, and to my amazement, the Board of Inspection and Survey recommended that no further installations of the *Somers'* type boiler be made until after complete service tests of that vessel—and this was with a war in the offing. The U.S. Gunboat, *Panay*, was sunk by the Japanese a month later, December 12, 1937.

The Board also recommended the elimination of all cruising turbines from 1500-ton destroyers. To this recommendation and the recommendation that a semi-enclosed feed system be substituted for the closed feed system installed on the *Somers*, I replied to the Secretary of the Navy as follows:[6]

[6] Letter from Chief of the Bureau of Engineering to Secretary of the Navy, via Chief of Naval Operations, DD381/58(11-4-Dr). Late in 1937 or early in 1938.

"4. The Bureau can only conclude that the weight of evidence to date both from laboratory test and from the operating experience which has been had with the divided wall, controlled superheat type of boiler, indicates that it is a sound and progressive advance in boiler design. This is especially true when it is considered that a further increase (to 850°F) in steam temperatures is necessary to fully utilize the 600-pound steam pressure to its full thermodynamic possibilities and makes the use of a controlled superheat boiler mandatory.

"5. . . . the Board [of Inspection and Survey] recommends the elimination of all cruising turbines from 1500-ton destroyers. The Board does not set forth the reason for its opinion, and it must be assumed that it considers a two-turbine arrangement more simple in operation and less liable to derangement by the reduction in units, and that these considerations outweigh all others. Operating returns from vessels similar in type to the *Somers*, but equipped with the two turbine arrangement favorably reported upon by the Board, in matters of repair and upkeep, have not convinced the Bureau that they are markedly superior to the *Somers* with an extra turbine. On the other hand, the gain in economy effected through the use of a cruising turbine in the *Somers'* installation is of such tremendous advantage from a strategical and supply point of view *in any orange [Japanese] war through which these vessels may have to fight,*[7] that it is worthwhile at this point to compare the return performances of the *Somers* with the *Porter*, a similar vessel as to size, type and age, but equipped with a two-turbine installation. Table I,[8] of course, affords a direct indication of the percentage of fuel economy which may be returned by alternatives of design either in cruising radius if that be desired, or in the utilization of excess fuel bunkerage for arms or armament if that be deemed most important.

"A similar comparison of the *Somers* (1850 tons) with

[7] The italics are the author's.

[8] Table I is omitted here because it has already appeared as Table II of this chapter.

the *Gridley*, a two-turbine arrangement 1500-ton destroyer, shows the *Somers* to be from 5 per cent to 30 per cent more economical in spite of her extra tonnage, a clear demonstration of her strategical superiority over the smaller and lighter vessel. While the Bureau will continue its attempts to develop an efficient competitive two-turbine design, it considers the evidence today overwhelmingly against the possibility that such a design can ever meet a three-turbine installation on an even basis.

"6. . . . The Board recommends that a semi-enclosed feed system be substituted for the closed feed system employed on the *Somers*, a recommendation they have carried forward through the twenty-six vessels of the *Mahan* class which have recently been placed in service. It considers such a system less vulnerable than a closed feed system.

"7. The Bureau can only consider that the Board is less alive to the absolute necessity of the removal of all oxygen from the boiler feed water when economizers are employed in naval boilers than the forces afloat have proved themselves to be. Commenting on the Board of Inspection and Survey's recommendation to remove the closed feed system from the *Dunlop* (DD 384), the Commander Destroyer Battle Force . . . states: 'This item is not recommended for accomplishment, and the Bureau has made numerous changes to improve operation and simplicity. It is considered that its use with economizers is imperative pending further service experience.' . . . the Commander Battle Force states: 'Commander Battle Force desires to emphasize that the closed feed system must not be removed from economizer equipped vessels until there is a system available that will furnish feed water free from the harmful content of oxygen. The system as now modified is a decided engineering advance; it is acceptable for service use; and its retention is recommended pending further developments.' . . . the Commander-in-Chief, U.S. Fleet concurs in the above recommendations that the closed feed system be retained . . . the Commander Destroyers Battle Force, recommends against the removal of the closed feed system from destroyers #364-379 and #384-393.

"8. The Bureau is well aware of the deficiencies of the closed feed system as it was installed in the many vessels which were laid down from 1933 on. However, it sought the ultimate in oxygen removal, and this can only be achieved in such a system. The semi-closed feed system is inherently less efficient in oxygen removal than the closed system, and it is not susceptible to improvement beyond its present capability. The closed feed system has such capabilities, and the Bureau has bent every resource at its disposal to the solution of a complicated and mechanical problem. Regardless of the results being obtained in existing service installations, it is already indicated that these can be modified to secure the result desired, and this will be done when practical experimentation has demonstrated a sure and economical method of remedying the known deficiencies. In the meantime, a reasonable and satisfactory success has already been attained. The Bureau cannot subscribe to the adoption of a principle which can only be made to operate moderately well when it is certain that further development and experience will ultimately produce the result which must be attained.

"9. In conclusion, the Bureau desires to express again its considered opinion that the *Somers'* engineering installation represents a distinct improvement over that installed in any other destroyer by any comparison of characteristics that can be made. It proposes to pursue the design to an adequate and proved conclusion in successive programs. It does not claim perfection in any plant so far completed, but feels very strongly that the continuous and progressive elimination of minor difficulties is crystallizing the design into a modern and up-to-date type of engineering plant as reliable, economical and space-saving as can be found in any Naval installation in the world today."

The issue was joined.

Immediately the lines of battle were drawn. Let us take

a look at the dramatis personae. My supporters in the Navy Department were Assistant Secretary Charles Edison and officers in my own Bureau below the grade of Captain, and such civilians as Lynn H. Korndorff, President of the Federal Shipbuilding and Drydock Company (U.S. Steel), William S. Newell, President of the Bath Iron Works Corporation, and William F. Gibbs of Gibbs and Cox, Inc. Arrayed against me in the Navy Department were the General Board, Admiral Thomas C. Hart, Chairman; the Board of Inspection and Survey, Rear Admiral Harry L. Brinser, Senior Member; the Bureau of Construction and Repair, Rear Admiral William G. Dubose; and, in civilian life, Bethlehem Shipbuilding Corporation, S. W. Wakeman, Vice President, Arthur B. Homer, Counsel, J. E. Burkhardt, Chief Engineer; New York Shipbuilding Corporation, J. F. Metten, Chief Engineer and later President; and Newport News Shipbuilding and Drydock Company, John F. Nichols, Chief Engineer. The Chief of Naval Operations, Admiral William D. Leahy, was apprehensive but neutral. Since no knowledge of engineering was required to join the opposition, its ranks were swelled by "straight" line officers and newspaper commentators. Considerable clamor arose in all quarters in regard to "high-pressure high-temperature"—most of it against. Even the usually cautious and reliable *New York Times* got off on the wrong foot.

None of the opposition realized that American industry under our system of regulated competitive capitalism had seized the technological supremacy of the world so long held by the British. They did not realize that this American success was due largely to the amazing progress in the art of engineering in our own country. They were completely lacking in appreciation of facts of the present and vision of the future.

I was by law the statutory head of engineering in the Navy, and I was determined I would be actually the Engineer-in-Chief no matter whom I had to take on, come what might.

Oxygen Determination

It was about 1936-1937 that we were having difficulty in determining the oxygen content of feed water. Samples too often became contaminated and there was no satisfactory equipment on the market to make rapid analysis. I asked the Naval Research Laboratory to invent a piece of equipment that could be used aboard ship for this purpose. They succeeded beyond expectation. The method and necessary apparatus were perfected by late 1939 and quite successful tests were made with the equipment at the Engineering Experiment Station boiler plant. By May 1940 work on the problem at the Naval Research Laboratory was completed. The report of the project was prepared in September 1940 but was not published until March 1941. The apparatus was portable and continuous in its recording. It revealed that the oxygen content of the feed water changed with the speed of the ship (throttle opening), whenever another feed pump was started up, whenever make-up feed was taken from a fresh water tank, and so on.

I was pestered by the Board of Inspection and Survey, which wanted me to make an oxygen survey of many ships while the Laboratory equipment was being built and at a time when I suspected the accuracy of all oxygen determinations on feed water. These suspicions were later confirmed when the laboratory equipment was in use. However, for the time being, I flatly refused to comply with the request of the Board of Inspection and Survey, stating my reasons. The Board appealed to Secretary Swanson, and I repeated my arguments to him. After considerable animated debate, I said to Mr. Swanson: "Mr. Secretary, there are about twelve Naval officers trying to be the Engineer-in-Chief of the Navy. My advice to you is to pick one that you have confidence in for Chief and fire the other eleven."

Admiral Leahy rejoined: "Mr. Secretary, Bowen is right. You have either got to support him or fire him."

There was much wisdom in Admiral Leahy's remark. It was a typical military reaction. After you have given a man a job to do, leave him alone. If he turns out to be no good, kick him out. The Secretary of the Navy is well supplied with Bureau Chiefs to do the Navy's work. Each Bureau Chief is, or should be, an expert and supreme in his technical field. He is recommended by the Secretary of the Navy and appointed by the President by and with the advice and consent of the Senate. The Secretary of the Navy is the chief coordinator of the Bureaus' activities and the court of last resort, so to speak. The technical staff of the Secretary are the Chiefs of the Bureaus, and he should not build up in his office another technical staff to review or supervise the Bureau Chiefs, nor should he use the General Board or any other agency as a staff for technical review. That is piling Ossa on Pelion and it is an organizational curse, typical of Washington but not peculiar to it. All this should be borne in mind when considering the case of the Board which recommended the merger of the Bureau of Engineering and the Bureau of Construction and Repair.[9]

Admiral Leahy's remarks ended the conference, and I did not make the ill-advised oxygen survey.

The Bath Change

The so-called "Bath Change" had to do with changing the design of destroyers building at the Bath Iron Works, DD 423 and 424, the *Gleaves* and *Niblack*, which were part of the 1938 destroyer program. This change was approved by the Secretary of the Navy in the spring of 1938.

It was an epochal change.

It resulted in the following eighteen, and succeeding, destroyers' having the same design:

1938 class, DD421-428, two—DD423 and 424
1939 class, DD429-436, all
1940 class, DD437-444, all

Never in the history of the Navy had so many vessels been planned and built to identical design. The machinery of these vessels incorporated all of the advanced features of Naval

[9] See p. 116, this chapter.

engineering which had been developed since 1933. Also, but not least, 600 p.s.i. and 850°F was adopted as standard for destroyers.

To put all this over required a lot of hard slugging.

Phase I of the Bath Change: Working plans for destroyers. This first phase deals with the designs for the 1938 class of destroyers, DDs 421-428, two of which were awarded to Bath and two to Bethlehem.

When the special specifications for DDs 421-428 were prepared, the same basic design features which had been originated in 1934 and 1935 were required. Among these were:

(a) steam pressure, 600 pounds
(b) steam temperature, 700°F
(c) high-speed turbines
(d) double reduction gears
(e) superheat control type boilers
(f) deaerating feed system

Bids were requested on DDs 421-424 in the summer of 1937 and six shipbuilders responded; namely, Federal, Bath, Bethlehem, United Shipyards, New York Shipbuilding Corporation, and Newport News. None of the bidders took any exception to the basic design features specified when they submitted their bids.

When the bids were opened it developed that the Bath Iron Works and Bethlehem Steel Corporation were low bidders. Their bids, including the cost of developing the details of the design and preparation of the working plans were as follows:

Bath —$5,075,000 for each of two ships
Bethlehem—$5,233,000 for each of two ships

It further developed from the breakdown of these bids that the estimated cost of developing the details of the design and preparing the working plans were as follows:

Bath —$800,000 ($400,000 for each ship)
Bethlehem—$400,000 ($200,000 for each ship)

Only one set of working plans was desired for this class of destroyer. Reducing the bid prices by the amount estimated by the bidders as the cost of preparing the plans gave the cost to the Government for the four ships as follows:

(a) Bethlehem to prepare plans:

Bethlehem	$5,233,000	for each of two ships
Bath	$4,675,000	for each of two ships
Total, 2 ships	$ 9,908,000	
Total, 4 ships	$19,816,000	

(b) Bath to prepare plans:

Bath	$5,075,000	for each of two ships
Bethlehem	$5,033,000	for each of two ships
Total, 2 ships	$10,108,000	
Total, 4 ships	$20,216,000	

The Bureau of Engineering was not convinced that the cost of the plans showed by Bethlehem in their bid represented a true cost of these plans. As can be seen by the above, a low estimate of plan costs by any bidder gives that bidder a definite advantage in the award of contracts. It is to the advantage of any shipbuilder to be charged with the development of plans since these can then be prepared in the sequence and method which best suits his individual requirements and in accordance with his custom. A following shipbuilder must adapt his methods of planning construction to the plan production of a leading contractor. It was the opinion of the Bureau of Engineering that the development of the details of design and the production of working plans for two destroyers cost much more than the $400,000 shown by Bethlehem in their bid. The estimate used by Bath, namely $800,000, was believed to be nearer to the actual cost of such work.

It should be noted at this point that Bethlehem's estimate of plan cost may have seemed to them to be reliable. I say this because at that time many of the shipbuilders, perhaps all, knew very little about accounting job order costs and estimating, and they knew nothing at all about budgeting funds, particularly on new construction. The preliminary report of the Special Committee on Investigation of the Munitions Industry certainly indicated that the shipbuilders, from their own testimonies, were not in a position to estimate completely costs or profits. I shall discuss this subject later in connection with the seizure and operation of the Los Angeles Shipbuilding Corporation. Then again, shipbuilders did differ widely in the amount of information recorded on plans.

The Bureau of Engineering was particularly anxious that Bath prepare the plans since their design agent, Gibbs and Cox, had had wide experience with the design specified for these ships. The counsel for Bethlehem Shipbuilding Corporation, Mr. Arthur B. Homer, conferred with me on this subject and was requested to change his bid so that an award could be made to Bath. This he refused to do and implied that, if Bath (Gibbs and Cox) prepared the design and plans, Bethlehem would not accept a contract. This left the Navy Department no alternative but to make the award on the basis of Bethlehem developing the details of the design and preparing the working plans.

Bethlehem had reason for not wanting to change its bid. A change in plan costs would not have changed the total bids. Bath could have prepared the plans and the Bureau could have continued its progressive course, a course in which Bethlehem had no confidence.

Phase II: Bethlehem proposes design changes in her contract for 1938 destroyers. Several months after the award of contracts for the construction of these 1938 class destroyers, Bethlehem proposed to the Bureau of Engineering that several basic changes be made in the design of these ships. In October 1937 officials of the Bethlehem Shipbuilding Corporation visited the Navy Department and requested that a design more in keeping with that previously followed by that company be authorized in lieu of that required by the specifications of their contract. Specifically, they proposed the following:

(a) steam conditions unchanged
(b) a high-speed high-pressure turbine driving through double reduction gears
(c) a low-speed low-pressure turbine similar to that previously used by this company
(d) integral superheater type boilers instead of the controlled superheat type specified

The only advance represented by these proposals, compared with the designs previously prepared by Bethlehem, was the adoption of a high-speed high-pressure turbine with double reduction gears in place of the low-speed high-pressure turbine driving through single reduction gears.

I was most reluctant to accept the contractor's proposals since I believed that they represented a definite lowering of the basic design standard which I had determined to pursue. However, I was faced with the fact that Bethlehem had a contract to design and build two destroyers, that they had no previous experience to guide them in carrying out the specified design, that they were definitely not in sympathy with the design policies of the Bureau, and that they were of the opinion that given an opportunity they could produce a destroyer of equal efficiency to those which incorporated the Bureau's ideas of design. I therefore granted, with the greatest reluctance, the request of the contractor and accepted their proposed changes in the design of DDs 421-428. In doing so I had no intention of perpetuating this bastard Bethlehem design, but I was willing to see a limited number of such ships constructed and to try them in service in competition with those built in accordance with the design policies of the Bureau of Engineering.[10]

Phase III: General Board changes military characteristics of hulls of 1939 destroyers—this change later made retroactive to 1938 destroyers. After the changes in machinery de-

[10] The *Benson* (DD 421) ran the trials for this Bethlehem class and failed to meet Guaranteed Fuel Oil Consumption.

sign to meet Bethlehem's proposal, the General Board of the Navy made certain changes in the military characteristics of the 1939 destroyer program (DDs 429-436) compared with those required for the 1938 destroyers. These changes did not affect the engineering installation but did have a considerable effect on the arrangement of the main deck and the hull. The Bureau of Construction and Repair, after conference with Bethlehem, decided to apply these new military characteristics to DDs 421-428 then under contract. This change was agreeable to the contractor, Bethlehem, and a delay of two months in the construction of the ships was authorized. As a result of this change in the military characteristics of these 1938 destroyers, together with the machinery change in them proposed by Bethlehem and previously authorized, the Navy Department was in the anomalous position of having a type of destroyer whose hull would be used for the 1939 program but whose machinery would not because the Bureau of Engineering had no intention of perpetuating the machinery design which Bethlehem had forced upon it.

Phase IV: Bureau of Engineering decides to require 850°F in 1939 destroyers. At this time (early spring of 1938), the Bureau of Engineering was engaged in the preparation of the special specifications for the 1939 destroyers (DDs 429-436). It was definitely determined, as the result of thorough investigation, to increase the steam temperature from 700°F to 850°F. It was also determined not to permit the building of any more destroyers of the design originated in part by Bethlehem in DDs 421-428 but to return to the all-American design. Meanwhile more experience had been gained with this American design in service, and its efficiency and ruggedness was more apparent than ever before. Twenty-six of the *Mahan* class, the first of the modern designs, had been completed and were in service, many of them operating with the Fleet. Reports from the Commander-in-Chief on the performance of this type of destroyer confirmed the belief that these vessels were rugged, reliable, and possessed a large cruising radius.

In addition, the *Somers* had conducted a shakedown cruise

of about 11,000 miles since her preliminary acceptance trials in the fall of 1937, in spite of the Navy's fuel-oil shortage. This extensive cruise had been directed by the Chief of Naval Operations in order to test the ship fully, since she had been criticized so violently by the Board of Inspection and Survey. During this cruise the vessel operated in both Arctic and tropical waters and was subjected to the most severe tests. Prior to the departure of the *Somers* on this shakedown cruise I had called her Commanding Officer, Commander James E. Maher, into my office to give him my orders. I told him to do everything he could think of to bust up the machinery of his ship. I didn't care what he did or how many times he went from full-speed ahead to full-speed astern, and vice versa, and if he could bust up the *Somers* I would take full responsibility for the orders I had given him. I doubt if anyone had ever received such orders before.

The Commanding Officer of the *Somers*, upon the completion of this cruise, appeared before the Assistant Secretary of the Navy, Mr. Edison, and the Chief of Naval Operations, Admiral Leahy, without my knowledge, and made an oral report on the performance of his ship. He told them that in accordance with my orders he had tried to bust up the machinery of the *Somers* and that it couldn't be done. From Commander Maher's report it became apparent that the points of ruggedness, reliability, ease of operation, and high efficiency were thoroughly substantiated and the dire forebodings of the Board of Inspection and Survey were completely unfounded.

Phase V: Two Bath boats, DDs 423-424 (1938 program) have their engineering design brought up to 1939 specifications. At about this time it became apparent that a delay in addition to the two months' delay authorized in the construction of DDs 421-428 was probable. This was due to a variety of causes, among them being the difficulty experienced by Bethlehem in producing a boiler design which was acceptable to the Bureau of Engineering. The changes in the hull further complicated the situation.

I came to the conclusion that the best solution would be to incorporate the 1939 engineering design in the two Bath boats and thus, in effect, make them the same design all over as the 1939 destroyer program. This decision would also expedite the delivery of the 1939 boats without unduly delaying the completion of the two 1938 Bath boats, DD 423, *Gleaves*, and DD 424, *Niblack*. The Bureau of Construction and Repair opposed this idea. Nevertheless, I recommended it to the Secretary of the Navy on 6 April 1938. This marked the final phase of the "Bath Change" and was approved by the Secretary of the Navy on 10 May 1938. The "Bath Change" was a monument to the untiring efforts and support of Assistant Secretary of the Navy Charles Edison, but the high-pressure high-temperature fight still had to be won.

Bids for 1939 destroyers. At the time the bids on the 1939 destroyers were requested, Bethlehem was designing, as amended, and building DDs 421-422 (1938 class). The steam condition for these ships was 600# 700°F. Table IV shows the major design features of the two classes of destroyers.

Bids *responsive to the specifications* were received from Bath, Federal and Bethlehem. These bids were as follows:

Adjusted Prices

Bath $5,067,000 for each of 2 ships
Federal $5,082,000 for each of 2 ships
Bethlehem $6,494,000 for each of 2 ships

In other words, Bath and Federal were low bidders on what the Bureau of Engineering wanted.

Bethlehem Submits Non-responsive Bid. In addition to the regular bids responsive to the specifications, Bethlehem submitted a bid not responsive to the specifications on the basis of reproducing DDs 421-422 (1938 class) then under construction at their plant.

This bid was: Bethlehem—$4,472,000 for each of 2 ships.

In spite of hell and high tide, Bethlehem seemed deter-

TABLE IV

	DDs 421-422 (Bethlehem) (1938 class)	DDs 423-424 (Gibbs & Cox) (1938 class & all 1939 class)
Steam conditions	600#, 700°F	615#, 850°F
Boilers	no superheat control economizers air casings desuperheater	superheat control economizers air casings no desuperheaters
Turbines	two-turbine arrangement consisting of a high-speed high-pressure turbine and a low-speed low-pressure turbine—astern turbine in one end of LP turbine	three-turbine arrangement consisting of a high-speed direct coupled cruising turbine, a high-speed high-pressure and low-pressure turbine—astern turbine in each end of LP turbine.
Reduction gears	double reduction on HP turbines—single reduction on LP turbine	double reduction on all turbines
Condensers	side supported	underneath type

mined to resist any change and to keep the boiler specialists and the gear specialists out of the shipbuilding business.[11]

Bath and Federal, with their subcontractors, would be indispensable if there should be an enormous expansion of wartime shipbuilding, which actually happened three years later.

Bethlehem's bid to build more of her bastard type DD 421 was not responsive to the advertisement for bids. To consider

[11] But see Appendix E, note 3.

this bid would have necessitated readvertising for bids on this type with consequent delays and the possibility that neither Bath nor Federal Ship would bid. It would have been a retreat on my part, and I was determined to put high-pressure high-temperature in all combatant types of steam powered vessels. The Bureau's design had met all tests at sea satisfactorily and, finally, the Parsons outfit might just as well know once and for all who was running the Navy's engineering.

In view of all these circumstances, I recommended to the Secretary of the Navy that the bid of Bethlehem to reproduce the DD 421 design not be considered and that award of the 1939 destroyers be made to Bath and Federal, the two low bidders responsive to the specifications.

The Secretary of the Navy approved my recommendation and made awards to Bath and Federal for the construction of two destroyers each. These awards were made in August 1938.

Thus the "Bath Change" was incorporated in the 1939 destroyer program. Standardization of ships' types had begun, and 600 p.s.i. and 850°F was standardized for destroyers.

High-Pressure High-Temperature in Battleships

North Carolina and Washington. The *North Carolina* and the *Washington* (BB 55 and 56) had been allocated on 24 June 1937 to the New York Navy Yard to build because the bids for these ships by private bidders were believed to be excessive. Shortly afterwards, and before the work on the ships had proceeded too far, I directed that the steam conditions be changed to 600 p.s.i. and 850°F from 600 p.s.i. and 700°F. Complete realization of the economy incidental to this change could not be realized because turbine steam nozzles were already made, steam piping had been procured, and so on, but most of the benefit could be and was realized.

Unfortunately, although this change in design, later alleged to be a change in military characteristics, was done with Assistant Secretary Edison's concurrence, the General Board (which drafts military characteristics) was not consulted, purely through an oversight. That didn't help a bit.

The design for the first two battleships of the new Naval Building Program was made at the Central Drafting Office in the New York Navy Yard, but they followed the new engineering which had been introduced in the destroyers.

Until the steam conditions were raised on the *North Carolina* and the *Washington*, the new engineering was restricted to destroyers of the Gibbs and Cox design and built at Bath, United Drydocks, and Federal Shipbuilding and Drydock Company. United Drydocks was soon acquired by Bethlehem, leaving only Bath Iron Works and Federal Ship (U.S. Steel) favoring high-pressure high-temperature and the new engineering.

It wouldn't be long before bids would be invited from private shipbuilders for four new battleships, and the "Big Three" (the old Parsons' yards), who would have to build three because there was room for only one in the Navy Yards (at Norfolk), were known to be bitterly opposed to high-pressure high-temperature.

The Edison Battleship Board

In January 1938 after I had changed the design of the battleships *Washington* and *North Carolina* to high-pressure high-temperature steam, Mr. Charles Edison, who was then Assistant Secretary of the Navy, convened a board consisting on naval officers and civilians who equally represented the proponents of and the objectors to high-pressure high-temperature steam.

A great deal of testimony was taken and the report of this battleship advisory board now reposes in National Archives and has never been given to the public. This report should represent the best cross-section of the naval mind prior to Pearl Harbor. Its release from security restrictions is vitally necessary to the historian of this period and the continuance of its confidential classification is absolutely absurd.

The attitude of my opposition was expressed succinctly by Mr. John F. Metten in a communication to this "Advisory

Board on Battleship Plans," dated 28 February 1938, as follows:

"The adoption of 600 boiler pressure for Naval work can not be justified for any material gain in economy unless the steam temperature is raised to approximately 850°F and this temperature should not be used for Naval work of any character.

". . . The type of closed feed system now specified for the battleships (BB 55 and BB 56) [*North Carolina* and *Washington*] is in my opinion undesirable. It has been installed in similar destroyer installations referred to, where it has been one of the most obvious sources of crowding the engine room and the complication of the feed system as mentioned in Naval Trial Reports. Equally effective lighter and simpler means of obtaining satisfactory deaeration of feed are available and should be employed.

". . . The type of divided furnace superheat control boilers specified for the battleships is considered undesirable. They have been installed for the first time in two destroyers that have just completed trials and must be regarded as experimental until they have had a year or two of continuous service operation with the Fleet.

"My objection to the proposed increase in pressure and temperature is largely from the complications that are necessary adjuncts and which result in disadvantages that are of more importance than any possible gains in other directions."[12]

The General Board gets into the act. On 18 October 1938 the Secretary of the Navy directed the General Board of the Navy "to investigate and report upon steam pressures and temperatures most advisable for use in combatant ships of new construction."

Bids were to be opened for more battleships (BBs 57-60) on 2 November 1938. It had been specified that these ships would be built to 600 p.s.i. and 850°F, but this had been

[12] See Appendix E, Battleship Advisory Board.

accomplished only after much maneuvering and after the usual controversy that attended every effort to improve naval engineering.

Originally the bids were called for in the usual manner. However, opposition to the original specifications was so bitter among the former Parsons' licensees, the "Big Three," that a Class II bid was announced in which the Government (Bureau of Engineering) would furnish the propelling machinery and the successful bidders would build the hull and install the machinery bought by the Government (Bureau of Engineering) from the same machinery builders who had supplied the modern machinery for the new destroyers. This Class II bid was originated because it was feared that the "Big Three," controlling the only yards in the country capable of building battleships, might not respond to the Class I bid and the Navy yards were building all the battleships they could handle at the time. This was a serious situation with international affairs steadily worsening. Chamberlain had just surrendered to Hitler at Munich, 30 September 1938, but my adversaries were too contentious to see the growing storm clouds.

Even so it was necessary to include in the contracts, at the suggestion of Bethlehem, a limited responsibility clause which expressed the idea that, since the contract included certain advanced features in engineering with which the contractors were not familiar, the Government would assume responsibility for any work involving additional expense which might be incurred by the contractor when it could be proved that this additional expense was the result of 850°F and was not due to defective workmanship or defective material.

In the case of Newport News Shipbuilding and Dry Dock Company, the Government actually purchased and supplied the propulsion machinery for the battleship they built under the 1938 program so that Newport News would not be liable for any such dreadful innovation as high-pressure high-temperature.

General Board hearings on high-pressure high-tempera-

ture. On 25 October 1938 the General Board of the Navy held extensive hearings at which witnesses representing both sides of the controversy testified. As is customary, the hearings were held by a subcommittee of the Board, but, contrary to custom, the presiding officer in charge of the hearings was not a member of the General Board but an Engineering Duty Officer, Rear Admiral Ormond L. Cox, head of the Naval Engineering Experiment Station, who was known to be hostile not only to high-pressure high-temperature but also to the Bureau's submarine policy, the policy which later enabled our submarines to make their splendid contribution to the war against Japan.

As Engineer-in-Chief of the Navy, I presented my argument in the following statement:

> "Statement of the Bureau of Engineering to the General Board in regard to the employment of steam of 850°F in the 1938 Battleships. [Undated, but shortly after 18 October 1938, probably 25 October 1938]

"1. The Bureau of Engineering selected 600# and 850°F as the standard steam condition for the battleships of the 1938 program and all other steam vessels for the following reasons:

"An inspection of the steam characteristics of power plants of the United States indicated that this combination of pressure and temperature was a popular one and the Bureau of Engineering reports practical experience to be gained from this source rather than from any other source or combination. 600# and 850°F is, of course, merely a step to 1200# and 850°F or possibly 1200# and 950°F. The results of any future increase of pressures and temperatures that may be recommended by the Bureau of Engineering will be dependent upon results obtained from the re-engined *Dahlgren* [World War I destroyer] which it is hoped will be operating at 1200# and 950°F early in February [1939]. Previous investigation by the Bureau of Engineering and the testimony of witnesses before the Battleship Advisory Board clearly indicated that the most ultra conservative did not consider that the introduction of steam at pressures up to 1200# constituted any Naval hazard.

"2. I feel, therefore, that I am warranted in continuing the remainder of my statement on the information which the Bureau of Engineering has which I believe proves to impartial observers that the advancement of steam temperatures to 850°F not only does not constitute a Naval hazard but institutes only those problems which are characteristic of any attempt that anyone might make in the field of progress. I realize, of course, . . . that there may be some argument as to the thermodynamic advantages to be obtained by this combination of 600# and 850°F. Of course, a complete analysis would require that this combination be examined alongside of almost any other combination of pressures and temperatures, but I am well aware that there are those who feel quite emphatically that almost the same thermal results could be obtained by a combination of 600# and 750°F. My answer to that is that all increase in efficiency in engineering . . . is obtained by . . . slight advances. The whole history of engineering is replete with instances where great results in the aggregate were only the result in detail of a pick-up of two or three per cent increase in efficiency here and there.[13]

"3. The Navy has successfully operated ships for some time now at 600# and 700°F. Many of our newer ships have been built gratuitously by the contractor to operate at 850°F. The first two ships in this category are the *Somers* and *Warrington*, and orders have already been issued by the Bureau of Engineering to gradually increase the temperature to 850°F in order that the Bureau may have early information of the operating and design difficulties, if any. In all ships of this type, it must be realized, we have as our servant for the first time in the Navy, controlled superheat. During the trials of the *Benham* a few days ago, it has been reported to me that the superheat did not vary more than one degree during a marked change of speed. The Board is well aware

[13] See Appendices C and D.

of the struggle that has always characterized every attempt to improve Naval engineering in the Navy since, in 1837, the first contract was let for a vessel to be propelled by steam as well as sails. It is perfectly natural that mankind should fear the unknown but, at the same time, the mastery of the unknown and great empires have been built by those who had no fear of the unknown. It is my firm belief that the importance of the inquiry now under prosecution by the General Board far transcends any mere question of whether our ships shall be built to 850°F or 750°F. I am convinced that your hearings will demonstrate that the raising of temperatures from 750°F to 850°F is not a hazardous step. I am equally convinced that whether the Board wishes to or not, it must weigh, in arriving at its conclusions, whether or not the Board is willing to take the responsibility of tempering the ardor of those who would judiciously and cautiously proceed into regions not already charted.

"4. The operating difficulties which will arise with 850°F must naturally be due to overstressed conditions during the usual course of operation and on account of the expansion of metal which is associated with this temperature, and also due to an ultimate deterioration of the metal due to this temperature.

"5. The fact that power plant engineering in the United States has successfully coped with the ordinary stresses which arise in their usual operations should be, I think, proof positive that the degree of expansion characteristic to this temperature is something which the art of engineering is at this time able to cope with.

"6. It is perfectly true that any ultimate deterioration, crystalline or otherwise, which might occur at this temperature has not been positively determined and will not be for some years hence. We are now discussing a matter whose generation is 20 years, and 20 years has not yet elapsed since the early installations in power plants.

"7. It is true, however, that the most elaborate laboratory methods have been used in an endeavor to anticipate any failure from this cause from metal.

"8. . . . .

"9. There may be those who have the feeling that plain ordinary operational difficulties will be encountered aboard ship . . . I, personally, do not feel that we shall encounter any such difficulties. . . .

"10. . . . .

"11. I would be derelict in my duty if I did not hereby call your attention to the fact that there is a great schism in the marine engineering profession of the United States. Some of the elements of this schism are so deep and so fundamental that, in my opinion, it is a vital necessity that the General Board shall consider them in their deliberations. For the sake of example, I will say that the manufacturers of power plant turbines are the General Electric Company, Westinghouse, and Allis-Chalmers. It is the practice of the so-called 'Big Three,' Bethlehem, New York Ship, and Newport News, to manufacture their own machinery although I have been assured by Newport News and by Bethlehem that they have no policy at all which prevents them from buying their main machinery when such action seems to them to be expedient. It is a practice of the 'Little Three' which has now diminished to the 'Little Two,' to operate, as far as Naval vessels are concerned, assembly plants. Bath and Federal (and formerly United Drydocks), prefer to buy their machinery. In view of the plant investment of the 'Big Three' it is only natural that they should prefer to make their own machinery; contrariwise is true of the 'Little Two.' . . .

"12. I will not, at this time, go into certain elements of British power plant practice because, while it has a bearing on the remarkable conservativeness and lack of foresight on the part of the British Navy, the General Board may not feel that it is strictly pertinent; however, I wish to assure the General Board that I have been furnished with a vital piece of correspondence between the Allis-Chalmers Company and Parsons, Limited. In this piece of correspondence Parsons, Limited, in giving a list of their outstanding installations from London to Rangoon and from Rangoon to Regina, states, all of these installations are characterized by approximately 600# and 800°F. In one of the last paragraphs of the letter from Parsons, Limited, to Allis-Chalmers, they state as follows:

> 'We are well aware that the power plant industry in the United States has advanced far beyond the limits of pressure and temperature that we have enumerated above, however, Parsons, Limited, wishes to assure you gentlemen that we are prepared, at any time, to build to those same limits of pressure and temperature as characterizes the most advanced installations in the United States.'

"13. In studies which may be presented later to the General Board, if the Board so desires, the Bureau will attempt to differentiate between those increases in economy which are due solely to increased turbine speeds and multiplicity of turbines, and those which are due solely to rise of temperature. . . .

"14. Again, speaking in general terms, it is a matter of fact that the machinery built into our new vessels and which is an adaptation of power plant practice is much more simple, rugged and efficient than that proposed by the shipbuilders who manufacture their own main propulsion machinery. The absolute and final decision as to this ruggedness and durability cannot be determined for several years but I am positive that no one could possibly inspect the sectional profile drawings of the two respective classes without coming to the same conclusion as I have—that power plant machinery design far excels the design of the shipbuilders in this feature. We have been very fortunate of late on account of the trials of the *Benham*, just completed at Rockland, to get a very good comparison between the two classes of machinery. The *Gridley* demonstrates a Bethlehem build and the *Benham* demonstrates a Westinghouse build. These ships are identical in everything from hull to steam conditions except that the machinery of the *Benham* is high-speed machinery. The results of these two trials are in the following table. [See Table III of this chapter]

"15. To summarize—there is no reason to suspect that operation at 850°F in a properly conditioned plant will offer any more development difficulties than those which have been continuously encountered by the Bureau of Engineering during its long history of attempting to improve power plants of our ships. There is no evidence which indicates that any unusual strains cannot be accommodated, that any unusual deterioration of metal will be set up, that there will be any such reduction in clearances as to later render parts ineffective, or that there is any unusual danger from the presence of autogenous ignition."

The attitude of the General Board was indicated in their communication to me as Engineer-in-Chief dated 15 November 1938:[14]

"1. In connection with the investigation of the use of higher steam pressures and temperatures, the General Board has become increasingly cognizant of the close dependence of the utilization of these pressures and temperatures upon the excellence of the design and manufacture of the machinery in which they are applied.

"2. In fact, it appears to the Board, that much machinery supplied to the Navy has not been entirely satisfactory regardless of the steam pressure and temperature in use.

"3. It is requested that the Board be furnished with a brief summary showing any unusual conditions that have developed on acceptance and final trials, and in service, of vessels of the heavy cruiser class and the destroyers beginning with the *Mahan* [363] class to date. This summary should include the name of builder, the designer and manufacturer of main propelling machinery and main auxiliaries, the casualties experienced with the above equipment, rejections, major repairs and replacements that have been necessary, with the steam pressure and temperature used in each case. This summary need be only in such detail as will allow the Board to identify the vessel, the manufacturer, and the building organization concerned, with the results obtained."

[14] Serial No. 1816.

The Board was entirely correct in its statement "that much machinery supplied to the Navy has not been entirely satisfactory regardless of the steam pressure and temperature in use."

No one knew that better than I, who had inherited all the problems from my many predecessors. For instance, reduction gears had to be renewed on many old cruisers; on so many in fact that it taxed the gear-making facilities of the country. The reduction gears of the new carriers, *Yorktown* and *Enterprise*, had to be replaced because in operation they produced in the engine room an unbearable noise. In fact, you couldn't even hear yourself talk. These gears had been built by a shipbuilder and not a gear manufacturing specialist. Other inherited problems I detailed in my memorandum to the Coordinator of Shipbuilding, Mr. Charles Edison (Assistant Secretary of the Navy), dated 24 October 1938.

The memorandum stated, in effect: two defective high-pressure turbine castings on the destroyer *Hull* would have to be replaced. These castings had been made for the builder, New York Navy Yard, by Bethlehem. Consequently, inspection of similar castings on three other destroyers had become necessary. Two high-pressure turbines on the destroyer *McCall* had been damaged on account of insufficient tip clearance. Increasing this tip clearance would emphasize still further the superior economy of turbines built by engine builders and not shipbuilders. The *McCall* was built by the San Francisco Union Plant of Bethlehem.

The contracts for castings to replace these defective castings represent the first time that the Navy had been successful in including gamma-ray examination of castings in the specifications of a contract.

The attitude of many "straight" line officers is revealed in extracts from a memorandum from Rear Admiral Joseph R. Defrees, Head of the Navy Yard Division and, as such, an assistant to Secretary Edison, as follows:

"7. There are two schools of thought on the high-temperature question. The more experienced seagoing line officers, who will be held responsible in Fleet action, favor the lower temperature side. Even the design engineers, including EDOs [Engineering Duty Only officers], are divided on the subject.

"8. The Chief of the Bureau of Engineering has not sold the high-temperature design to operating Naval personnel. *They question the advisability of gambling with National defense by installing experimental and unproven engine designs which are not required to meet the military characteristics laid down by the General Board.*[15]

"9. The most important factors are reliability, simplicity, and ease of operation, all of which are decreased by increased pressures and temperatures."

This memorandum, as quoted, sheds considerable light on what was in the minds of some of "the more experienced seagoing line officers" in their opposition, not only to high-pressure high-temperature but also to the multi-cylinder diesel electric drive for submarines.

The General Board was the last word on "military characteristics," but the Engineer-in-Chief was the last word, or should have been, on engineering progress. It was my duty, not the General Board's, to be positive that the billions we were to pour into the future building program should not be invested in obsolete and outmoded machinery. In attempting to dictate engineering design, the General Board not only exceeded its authority under the specious plea that design in engineering is a military characteristic, but it invaded a highly technical field in which it was not competent to function and it also attempted to assume responsibilities which are the Engineer-in-Chief's alone. As a matter of fact, the Secretary of the Navy never should have referred this controversy to the General Board in the first place. This is one of the many instances where the General Board has been used to get the Secretary of the Navy off the hook.

The Board of Inspection and Survey, whose duty it is among other things to see that the terms of the contract and

[15] Italics added.

the guarantees are met with respect to new vessels, fell into the same fallacious reasoning and never ceased attempting to arrogate to itself design functions which were my sole responsibility as the Engineer-in-Chief of the Navy.

It is also noticeable, in the last paragraph of the quoted memorandum from one of "the more experienced seagoing line officers," that among the most important factors in military characteristics, economy, i.e. cruising radius, is conspicuous by its absence.

The Bureau of Engineering was convinced that to continue to install old-fashioned, uneconomical machinery really would be "gambling with the National defense," would be a wicked waste of money, and would not be preparing our Navy for the long reaches of the Pacific where war sooner or later was inevitable. We strove to the utmost in all classes of vessels to increase their economy or cruising radius for irrepressible war with Japan.

There was, unquestionably, in the minds of those who stretch the expression "military characteristics" to include everything under the sun, deep resentment that the Bureau of Engineering was moving heaven and earth to increase the cruising radius of our new ships above that prescribed by the General Board. They unquestionably felt that in doing so the Bureau was invading the field of the General Board. World War II proved conclusively that it was fortunate that someone was worried about cruising radius, even if it might mean invading the field of the General Board.

The Royal Navy was cold to our engineering developments. The Engineer-in-Chief of that Navy was repeatedly quoted by responsible individuals to have said in conversation with visiting Americans that our advances in steam conditions were not warranted by the state of metallurgy, that our Bureau of Engineering had been "sold down the river by the same big electric manufacturing companies who put over electric drive on the Bureau of Engineering," and "let the Americans and Germans do it, and if it succeeds we will copy them." The fact that the Royal Navy did not pioneer in high-pressure high-temperature and would not even follow us was also an element in the minds of the opposition.

I felt that the scope of the hearing had been greatly broadened in view of the Board's letter of 15 November 1938, previously quoted, and that it was incumbent upon me to submit a further statement with respect to general engineering matters. The following extract is quoted from my memorandum dated 30 November 1938:

"13. From certain questions that have been asked and statements that have been made it appears proper to give consideration to the question of economy. There apparently has been some reversal of opinion on this subject. Prior to the entry of the *Mahan* class into the Fleet it was my understanding that the most vital defect in our Light Forces was their lack of radius of action. The situation was so acute that the first maneuvers in a General Fleet Problem was to steam about 1,000 miles to sea and refuel all the destroyers. During that period many officers in high command complained bitterly about this vital deficiency. More recently during the last maneuvers in the mid-Pacific, the long cruising radius of the new destroyers, particularly the *Mahan* class, was most favorably commented on by Force Commanders and others. The inability of the wartime [World War I] destroyers to steam from San Diego to Honolulu, to participate in problems, and to return to Honolulu without refueling was commented on in contrast with the new destroyers of long cruising radii which were able to perform such evolutions. The long cruising radii and the high speeds of modern destroyers have been favorably commented on also by Aviation Force Commanders who state that this is the first time they have felt that patrol planes were furnished with adequate plane guard. In engineering circles at the time there was no extreme to which active consideration was not given for a remedy. One such was the employment of diesel engines for cruising purposes such as was being developed and tried in German light cruisers. . . . The carrying of large excess quantities of fuel also had to be ruled out on account of limiting displacements. . . . The development of high-speed turbines and double reduction gears has provided a very effective solution of this important military characteristic of radius. The statement has been made that nothing should be sacrificed for economy. I cannot subscribe to such a theory. . . .

"14. It has been suggested that new developments be tried out on one ship of a class as an experimental unit. However, when it is considered that three to five years are spent in the construction of a ship, perhaps another year to clear up defects, the time element becomes important in that four to six years of construction have passed before the experimental results are known. Such a course would appear to deprive the Navy of the best that can currently be obtained."

In summary I pointed out that a well conceived improvement in propelling machinery had been accomplished and would be continued, that much criticism of existing machinery was directed really at the *Mahan* class, that the admitted congestion on the *Mahan* was not wholly bad since the weight and space thus saved had been utilized by the Bureau of Ordnance, that the question of space and accessibility no longer existed since it had been corrected in later ships, that no criticism could be leveled against the *Mahan's* machinery, that the superiority of the *Somers* and *Benham* over the *Porter* and *Gridley* vindicated the modern machinery, and "*that the matter of economy of operation and therefore cruising radius is a vital military characteristic of such importance that it cannot be lightly treated and certainly cannot be completely disregarded for the sake of comfort and convenience.*"[16]

"16. The attempt of the Bureau of Engineering to keep its installations in line with the progress of engineering has at various times met with astonishing and unreasonable opposition in the Navy Department. In 1869 in Navy Department General Orders we find: 'Hereafter all vessels to the Navy will be fitted with full sail power, the exception of this

[16] Italics added.

will be tugs and dispatch vessels not fitted with sails.' Shortly after, the issuance of a General Order was contemplated warning Commanding Officers that any unusual expenditure of fuel would be charged to their personal accounts, the idea being, of course, to discourage any use and development of main engines of the war vessels in existence at that time. It was 1875 before the Bureau of Engineering was able to use 4-bladed instead of 2-bladed propellers. The vacillating policy by the Bureau in connection with the introduction of turbines and of fuel oil in place of coal would seem to indicate some such lack of harmony in the Navy Department. . . . The attempts of the Bureau of Engineering to progress have, as I have indicated above, been retarded from time to time by outside influences. The trend in marine engineering today is definitely indicated—it is toward higher pressures and higher temperatures—the near ultimate being 1200# and 950°F. The new construction including the battleships have been laid out for 600# and 850°F which is a moderate step along the lines which the trend indicates. Any reversal of this policy of the Bureau of Engineering by higher authority merely means that battleships which take 4 to 5 years to build will have machinery ten years behind the times when they are commissioned."

On 1 December 1938, and after the hearings were concluded, the General Board stated in a communication to the Secretary of the Navy:[17]

"1. . . . .

"2. The General Board is of the opinion that the advance to steam at 600 pounds and 850 degrees has been made much too rapidly in view of the lack of practical experience of these figures afloat. *The important factor of economy has evidently been overweighted and its stress has tended to complexity in some features of design that involve considerable risk in reliability and in difficulty of repair and maintenance.*[18] It appears that sufficient advantage has not been taken of op-

[17] G.B. No. 420-13 (Serial No. 1816a), dated 1 December 1938.

[18] The italics are the author's.

portunities to test this type of installation at the higher temperature during the past year or more on several Navy ships, so that there is no data now available upon which to base definite conclusions or recommendations. However, the evidence adduced is largely opinion rather than fact and therefore does not justify the discontinuance of installations of machinery at the higher figures of pressure and temperature for ships now under contract.

"3. . . . .

"4. . . . the General Board recommends that intensive accelerated comparative tests be conducted as rapidly as possible with destroyers having both types of pressures and temperature plants so that practical data may be available upon which to base decisions as to future construction."

All of this with Pearl Harbor only three years away!

In the following memorandum[19] to the Assistant Secretary of the Navy dated 12 December 1938, I commented on the above report of the General Board, reference (a) of the quotation below, and a supplementary report of the General Board,[20] reference (b) of the quotation below.

"1. Referenced letters report the General Board's findings on the subject of steam conditions for new construction. While no clear directive is indicated in these reports I feel that many inaccuracies in them might lead to erroneous conclusions and action and therefore take this opportunity to submit certain comments upon them.

"2. Referring to reference (a), I will comment only in a brief and general way reserving more detailed comment for the supplementary report, reference (b). The Board states that the question of steam conditions resolves itself into the use of steam either at approximately 450 pounds and 700°F or 600 pounds and 850°F. . . . Actually the Navy has in successful operation many ships at 600 pounds pressure and 700°F temperature so that the question of pressure is no

[19] Bureau Engineering letter F.S/S40(12-12-C) Confidential, dated 12 December 1938.

[20] G.B. No. 420-13 (Serial No. 1816b) Confidential, dated 1 December 1938.

longer a matter of opinion or experiment—it is an accomplished fact. So far, therefore, as any question of experimentation is concerned the matter resolves itself into a question of 700°F versus 850°F temperature. Contrary to the Board's report, operation with 850°F temperature has recently begun and is continuing in the *Somers* and *Warrington.* Earlier experience with 850°F operation was not undertaken for reasons that will be explained in subsequent paragraphs. *In the matter of economy it has been reported to the Board that no serious sacrifices have been made on this score and it is incomprehensible how a military characteristic of such vital importance can be treated so lightly by the Board.*[21] In this connection and should a reduction in steam conditions be seriously considered, it is important to note that in January bids will be invited for four cruisers of moderate speed (CL 51-54). The characteristics called for by the General Board include a speed of 32 knots and a cruising radius of 8,500 miles. If the steam conditions, 600 pounds and 850°F presently specified for these vessels, were reduced to 450 pounds and 750°F, the speed would be reduced to 30.6 knots and the cruising radius to 7,775 miles.

"3. The General Board's supplemental report, reference (b) is in considerably more detail. The following comments refer to corresponding paragraphs in the report.

"*Paragraph 2.*

"Board report—'The Board has evaluated such facts as are available. The comparative data is meager. At high levels of pressure and temperature the only data available to the Board are those from steam plants on shore where weight and space do not enter the problem and in which the conditions and the objectives are markedly different than in combatant ships. There is no shipboard service experience in the Navy with steam at 600 pounds and 850 degrees. Very recent orders have been issued to two destroyers to operate at this pressure and temperature. *One class of destroyers which have now had considerable service were so built that steam*

[21] Italics added.

at 850 degrees could have been used with certain minor alterations. Had one or more of them been so employed in extensive running under service conditions, there would now be available concrete data on the subject. It is most unfortunate that such steps were not taken.'[22]

"Bureau comment—The Navy has had very considerable experience in a number of ships with steam pressure of 600 pounds which has worked out entirely satisfactorily. To this extent the General Board's statement is incorrect. Operation in service at 600 pounds, 850°F in both *Somers* and *Warrington* was reported to the General Board in the hearing of November 30th. While such service had extended only up to 22% power at that date, the General Board's statement in this respect is incorrect. The General Board infers that reference to power plant operation with high steam conditions is not proper on account of the widely differing conditions of space and weight. There are many problems in the employment of any set of steam conditions, ashore and afloat, which are the same. It would appear to be only a matter of common sense to make use of the developments made in shore plants so far as possible in the development of Naval machinery. . . .

"The General Board states that it would have been possible to obtain earlier experience with operation at 850°F and that it was most unfortunate that such steps were not taken. This refers to the *Mahan* class. The *Mahan* class, commissioned beginning 1935, was constructed to operate at 850°F provided rather extensive changes were made in the superheaters. But, as reported to the General Board, the *Mahan* class already presented two important steps in advance in machinery; namely, an increase in temperature from about 600 to 700°F and the introduction of high-speed economical machinery. It would have been a violation of conservative progress and good judgment to have further complicated that program with more development. The tenor of the General Board's report is certainly one of conservatism and it is hard to see the consistency of questioning the

[22] Italics added.

Bureau's procedure now in a matter which they think should have been done three years ago. I particularly object to the use, by the General Board, of the word 'unfortunate' in this connection. No further development can possibly occur in construction at 600#, 850°F which would justify such an apprehensive term."

As a matter of fact, it was my predecessor who got cold feet and changed the steam temperature of the *Mahan* class from 850°F to 700°F, probably to avoid getting into the controversial mess and bitterness which I got into with high-pressure high-temperature.

"*Paragraph 4. Space Requirements.*

"Board report—'With "high" steam the work is done with a less amount of water and in theory everything is smaller than with "low" steam. That condition is not borne out in practice. Turbines are smaller, boilers may be and pipes are smaller with high steam. But reduction gears tend to be larger, lagging is somewhat thicker, steam pipes are longer. There is also a tendency to use more auxiliary devices and attachments and congestion appears to be actually somewhat greater. There is no definite proof that either type of machinery requires less space than the other.'

"Bureau comment—The General Board's statement that reduction gears tend to be larger and that there is a tendency to use more auxiliary devices and attachments, and congestion with high steam conditions, is definitely in error. These features are not affected by the level of steam conditions. That there is no definite proof that this type of machinery requires less space than the other is a fair statement.

"*Paragraph 5. Weight.*

"Board report—'In this comparison also there is no well defined difference but such as is apparent is in favor of low steam, the recent installations being lighter than contemporary high steam plants.'

"Bureau comment—The General Board's statement in respect to weight is incorrect and unfair and totally disregards the statement of the Chief of the Bureau of Engineering made to the General Board in the hearings of 30 November. . . . As pointed out in the hearings, however, a direct comparison of machinery weights in those two ships (*Somers* and *Porter*) is quite unfair in view of the wide difference in their performance. It is a comparison between a ship (*Somers*) of 38.85 knots speed and cruising radius of 10,540 miles against a ship (*Porter*) of 37.17 knots speed and cruising radius of 8,710 miles."

Here I cited comparative figures for the *Somers* and *Porter* as indicated in Table II and the *Gridley* and *Benham* as in Table III.

"*Paragraph 6. Cost.*

". . . .

"*Paragraph 7. Production Factors.*

"Board report—'Both high and low steam plants can be built, probably in ample quantity. In general, the producers of high steam machinery components have greater design and research facilities and their production has been carried on more continuously. The makers of low steam machinery components have had most of the general experience in producing naval engineering equipment and are better conversant with the conditions under which it operates. Either class of manufacturers is more likely to produce efficiently the type of plant with which it is familiar.'

"Bureau comment—The production capacity of the manufacturers of machinery, namely, General Electric, Westinghouse, and Allis-Chalmers, is immeasurably greater than that of the shipbuilders. None of the shipbuilders has a capacity appreciably beyond that required for machinery for the ships they build. On the other hand the engine manufacturers are in a position to turn out machinery in large quantities. It is also very important to note that the machinery manufacturers have extensive design and research facilities which are a prerequisite to satisfactory and continually improving machinery. These engine manufacturers are competent and experienced in the design and construction of high-speed machinery. The shipbuilders are not and their opinions and testimony before the General Board are undoubtedly influenced by that fact. Both General Electric and Westinghouse have had wide experience with marine installations and Allis-Chalmers has had some.

"*Paragraph 8. Fuel Economy.*

"Board report—'The advantage is definitely with high steam, which is the reason for its adoption in power plants where economy is of greatest importance. The theory which applies is that high steam increases fuel economy 10 per cent. Claims have been made on the basis of recent destroyers' trials that a general increase in economy of 15 to 20 per cent is realized. It is found that much of such increase is the result of added installations and complexities not necessarily connected with pressure and temperature. With machinery that is comparable on the same basis throughout and using the most probable circumstances in war, about 6 per cent increase in economy is probably the most that can be expected with high steam.'

"Bureau comment—The statement of the General Board is incomplete and inaccurate. The actual gain in fuel economy to be obtained with high steam conditions as compared to low steam conditions will range from about 6 per cent at cruising speeds to a slightly larger figure at high power. If high-speed turbines and double reduction gears are included along with the higher temperatures the increase is in the order of 15 to 25 per cent. This latter figure is really the one which should be used in making the comparison between high and low steam because of the fact that low-speed machinery is conveniently applied to the low steam conditions whereas high-speed machinery naturally follows the high steam conditions. It is in error to state that much of such increase is the result of added installations and complexities. The steam conditions have practically nothing to do with the complexities of modern installations. Trial trip figures speak for themselves in this matter of economy."

Then the trial trip data included earlier in this chapter were quoted. There is no need to quote the remaining six paragraphs of this long letter. The General Board and I were at loggerheads on all matters discussed.

The Board stated that there was some advantage in low-pressure steam plants, that high-pressure plants are more complex and therefore have a greater chance of operational error and demand more numerous and more efficient personnel. The Board objected to controlled superheat both per se and also because it was "an essential feature of 850 degree steam plants"; it worried about maintenance cost of the new plants, and greater temperature changes.

The Board felt that: "There have been errors of design in many plants due to giving far too much weight to fuel economy and thereby complicating production, operation and maintenance," and another gem: "Some machinery design shows far too little realization that the ships are built for combat and much too little foresight of what can happen from enemy action or lapses of personnel under fire," and again: "Too little attention has been paid to necessities of accessibilities for maintenance, and even examination, by designers who could not have had Naval conditions of service sufficiently in mind." The Board wound up by recommending that "intensive" comparative tests of the *Somers* and *Porter* classes be held in order to arrive at definite conclusions as to points in dispute between me and the General Board.

For my part, I rebutted the charges of the General Board aimed at my engineering judgment and also stated: "In conclusion I wish to invite the attention of the Secretary to the fact that the precept under which the General Board began its hearings had to do exclusively with steam conditions of pressure and temperature presently specified by the Bureau of Engineering for new construction. . . . Subsequently it became quite evident that the General Board had embarked on a fishing expedition in which many features of machinery design were being investigated. Some of the big [three] shipbuilders, who are unable to produce [the] most modern machinery, were doubtless very helpful with criticism. . . ."

I was never given an opportunity to confront and cross-examine my accusers, nor was I ever allowed to read their testimony.

Thus closed a most discreditable chapter in American shipbuilding and American industry, and it reflects no credit on the Navy Department as a whole.[23]

At a later date I was investigated by the General Board under the charge that I had departed from the military characteristics of destroyers as laid down by the General Board in that I had designed and produced destroyers which exceeded the speed and cruising radius laid down by the General Board. Needless to say, these hearings broke down in merriment and confusion.

High-pressure and high-temperature were installed in the *Atlanta* class of light cruisers, thus marking the entrance of the new engineering into the cruiser type of vessels.

Attempts to introduce it into carriers then planned or building would have resulted in some delay in the completion of those vessels and the White House was unwilling to accept the delay.

High-pressure high-temperature formed the background of propulsion engineering during World War II. According to Vice Admiral Earle W. Mills, later Chief of the Bureau of Ships, our operations in the Pacific would not have been possible without it.

Certain shipbuilders, as has been related, were forever quoting Royal Navy practice in order to prove that I was wrong in pushing high-pressure high-temperature and the newly designed auxiliary machinery. In order to find how the Royal Navy made out with the low-speed uneconomical Parsons machinery and small cruising radius, let us look at the record.

In a splendid book entitled *The Bismarck Episode*, by Captain Russell Grenfell, RN,[24] which all decadent shipbuilders should read, I find:

"The two principal British weaknesses—one strategical and the other technical—revealed by the operation related to fuelling and gun-mountings. The British effort to destroy the *Bismarck* very nearly came to grief on the rock of fuel shortage. The British fleet as an integrated force was quite

[23] See Appendix J for a similar episode, the "electric drive fight."

[24] The Macmillan Company, 1949, p. 198.

incapable of conducting a prolonged chase. Ships were soon dropping out by ones, twos, threes, and fours, because their fuel supplies did not enable them to follow the enemy any farther; and we have seen how nearly Sir John Tovey had to abandon the pursuit altogether on the very eve of success.

"The fact is that the Admiralty had consistently omitted, for many years, to give serious attention to the question of fuelling at sea. Had fast tankers formed a normal part of a seagoing 'task force,' Sir John Tovey's ships could have oiled at sea and so have pressed on after the enemy without the increasing anxiety that they felt and the careful rationing of their speed that they were compelled to practice.

"That no such organization existed in the British Navy can probably be ascribed to its two-century predominance among the fleets of the world having given it a copious supply of fuelling bases in every ocean. In consequence, British naval officers had become unduly 'base-minded,' the shore stocks of fuel at one of Britain's numerous naval bases being uppermost in their thoughts when any question of mobility arose. The American Navy, growing to maturity at a time when most of the outside fuelling bases were in other hands, had long been accustomed to take its fuel and stores about with it.[25] Later in the war, when the British fleet went to the Pacific where its own established bases had fallen into Japanese hands, it also had to adopt the American system and organize floating replenishments of every kind."

There is no reason why Captain Grenfell should have known about the desperate fight which a few of us waged in Washington to increase the steaming radius of our own naval vessels through increased economy and on account of the enormous distances in the Pacific. But the two top Admirals in the search and destruction of the *Bismarck* have expressed their opinions on this matter.

Admiral Sir James Somerville, famous for his exploits in the Mediterranean, who joined the *Bismarck* search with

[25] In view of the great distances in the Pacific and the small cruising radii of many of our combatant ships, the U.S. Navy was compelled to develop refuelling at sea—H.G.B.

his Force H, stated to me during the war that the superiority of the machinery in our ships in comparison with those of the Royal Navy was great indeed and he wished that the British ships had our machinery.

Admiral Sir John Tovey, Commander-in-Chief of the Home Fleet, who led the search in his flagship, *King George V*, was said to have been much exercised over the fact that our *Washington*, a contemporary of the *King George V*, was about thirty per cent more economical than his flagship.

On one cruise at sea, Tovey was dumbfounded to find out that, operating under similar conditions, the *Washington* required several hundred tons less fuel than the *King George V* upon returning to port to refuel.

In port the *Washington* was likewise more economical, notwithstanding the fact that the British use the "water boat" while our ships are required to distill their own fresh water for make-up feed and for drinking purposes.

The *King George V* burned 39% more fuel at low speeds than the *Washington*, and the *Washington* maintained a fuel superiority up to high speeds. Furthermore, the cruising radius of the *Washington* was double that of the *King George V* at low speeds and very nearly twenty per cent greater at high speeds.

The *North Carolina*, sister ship to the *Washington*, exceeded her designed speed of 26 knots by 2 knots. The *North Carolina* and the *Washington*, it will be remembered, were the first battleships to have high-pressure high-temperature and the new engineering.

In April 1941 I received a letter from a Naval Attache[26] of the American Embassy in London. He stated in part: "On the whole we are many years ahead of them [the British] technically. This is true in practically every phase of naval engineering. Many of them look with considerable awe upon our modern machinery installations such as the *Gleaves* [product of the Bath Change] design. . . .

"I also now understand much better than I did before why

[26] Commander, now Rear Admiral, Paul F. Lee, USN (Ret.).

our old line shipbuilders [the "Big Three"] insisted on attempting to design as well as manufacture most of the machinery used in the ships they build. This practice is quite universal here in England insofar as boilers, turbines, and reducing gears are concerned. I have visited every shipyard on the Clyde, and everyone has his finger in the design of all three of these important units. In the case of turbines, they all consult with Parsons and he in fact controls the design. Insofar as I can see, he has little competition in this matter and this probably explains the lack of advance which has been made in turbine propulsion here. The admiralty specifications are very restrictive and they apparently have not kept abreast with the developments made in shore plants. This latter they now realize as having been a serious mistake and I think they will follow the example which you set back home from now on.

"It would do certain of our people at home a great deal of good if they could sit in on some of the discussions which I have about the cruising radius of ships. This is a very serious question with the British and is having a marked effect upon their naval operations. Even in their newest ships the fuel consumption is at least 50% higher and in some cases almost 100% higher than we have in our modern designs. Due to war conditions the normal peacetime cruising radius of their ships has been reduced by as much as 50% in some classes. This latter, combined with the poor fuel economy has given their ships a comparatively short cruising radius. They are now fully alive to the mistakes they made in their pre-war designs, and I feel confident that for the designs which they prepare from now on they will make very great advances both in the steam conditions and in the type of machinery which they install.

"The old argument that a low-pressure low-temperature design would reduce the complexity of the installation is not borne out by the British ships I have visited. Their plants are as complex as are ours and in practically all cases the access to the machinery is inferior to that of our own ships. I would give a great deal if some of the critics of the modern machinery plants, which have been used there at home, could compare one of those plants to those installed in British ships built about the same time."

Overweight Destroyers

One of the controversies which followed in the wake of high-pressure high-temperature was the case of the overweight destroyers.[27] It was discovered on trial trips that one of our representative destroyers, when traveling at full speed in the trough of the sea with a stiff wind blowing and fuel-oil tanks nearly empty, did not respond quickly to hard-over-helm. Immediately there was raised a great furor in the press, but the actual reasons for the behavior under all the adverse conditions and conditions which might never be encountered in service are easy to understand.

We were laying down a new type of destroyer each year. When a ship is completed she is given what is called an inclining experiment in order to determine her metacentric height, that is, the distance between the center of gravity and the center of buoyancy. It was taking so long to build a new type of destroyer that at least two more classes were laid down before a ship was ready for her inclining experiment. The Bureau of Construction and Repair, the Bureau of Ordnance, and the Bureau of Engineering were very naturally trying to include in each type all of the things under their cognizance which they believed would improve the fighting quality of our destroyers, and, in the case of the so-called overweight destroyers, each of these three bureaus had exceeded its original weight estimates. There was one thing the Bureau of Construction and Repair, the former naval

[27] Criticism of design of naval vessels is not unique in our naval annals. There was a Congressional Investigation in February 1908 to investigate charges made by the crusading Admiral William S. Sims, USN, that the side armor of our battleships was mostly below the load water line, that the handling rooms of the turrets were not safe and were contributing to powder explosions in turrets, etc. The Admiral directed his fire principally against the Bureau of Construction and Repair and its Chief Constructor, Washington L. Capps, although the Bureau of Ordnance did not escape unscathed. (*Admiral Sims*, by Elting E. Morison, Houghton Mifflin, 1942, chapter 12.)

architectural bureau, did not include in its design at this time. That was a factor of safety in the metacentric height to anticipate increasing displacement. No engineer would think of building a piece of machinery without including in it a factor of safety. No one in the Navy at that time seems to have understood that progress means an increase in displacement.[28] Of course, the uninitiated and the uninformed charged the overweight destroyers to high-pressure high-temperature. There were some who would have blamed high-pressure high-temperature for bad eggs for breakfast.

In order to be on the safe side, all suspect destroyers had 50 tons of pig lead located in their bottoms.

Before and after Pearl Harbor, when I was Officer-in-Charge of the Federal Shipbuilding and Drydock Company at Kearny, New Jersey, Mr. William Francis Gibbs and I worked out a plan whereby we believed that sufficient improvements could be made in the destroyer hulls and their fabrication to eliminate the 50 tons of lead and, at the same time, ultimately expedite the completion dates of destroyers. For some reason Secretary Knox and the Bureau of Ships disapproved the initiation of the so-obvious improvement.

The Amalgamation of the Bureau of Engineering and the Bureau of Construction & Repair

It was my original intention when I came to this subject to treat it casually as a sort of end product of the high-pressure high-temperature steam fight, but after my attention was called to a recent allusion to this subject in print, a more detailed treatment seems to be demanded. So let's dish it out.

In his book, *The Navy and the Industrial Mobilization in World War II*,[29] Robert H. Connery states: "The Secre-

[28] As a matter of fact, very little attention was being given to metacentric height. Shortly after my attention was called to the subject I happened to be in the New York Navy Yard where I broached the subject to the Commandant and the Manager. Neither could tell me, off-hand, what the designed metacentric height was of the new battleships they were building.

[29] Princeton University Press, 1951.

tary's [Edison's] ideas on reorganization had grown out of his study of naval organization while considering the proposed merger of the Bureau of Construction and Repair and the Bureau of Steam Engineering.[30] It had been apparent for some time prior to 1939, he said, that there was a lack of cooperation between these two bureaus—primarily a result of personalities, rather than organization. . . ." Let us examine what Secretary Edison did state in his memorandum:[31]

"Early in the Spring of 1939 there became apparent a lack of cooperation between the Bureau of Construction and Repair and the Bureau of Engineering. As this lack of cooperation seemed to be based rather on personalities than on organization, I made an endeavor to make certain changes but was prevented from doing so because the Chief of Naval Operations and the Director of Shore Establishments advised the Secretary of the Navy to the contrary. This clash between the two principal shipbuilding Bureaus was brought to the surface and to my attention through a difference of opinion concerning the use of higher pressure and temperature steam conditions in the main machinery installations in the new ships.[32]

"Later on in the Spring of 1939 I received unofficial word that one of the new destroyers had failed to successfully pass her inclining test, indicating a deficiency in her reserve stability.[33] This matter came officially and formally to my attention in July 1939.[34] An investigation of this situation indicated that the stability calculations are the responsibility of the Bureau of Construction & Repair. This investigation

[30] The name of this Bureau was changed to Bureau of Engineering on 4 June 1920.

[31] Memorandum, Secretary Charles Edison to file: "History of Reorganization Effort," 16 December 1939.

[32] The Chief of the Bureau of Construction & Repair had no responsibility, statutory or otherwise, for steam conditions.

[33] This means that the actual metacentric height was appreciably smaller than the designed metacentric height.

[34] See p. 118 of this chapter. The records showed that on all new destroyers the actual metacentric height had been steadily diminishing compared to the designed metacentric height which in itself had been steadily reduced.

further brought out the fact that weights and centers of gravity of material supplied to this ship by the Bureau of Engineering and the Bureau of Ordnance had been changed somewhat from the original estimates of weight and center of gravity submitted to the Bureau of Construction & Repair,[35] upon which the latter Bureau based its stability calculations. Here is a glaring[36] example of the lack of coordination and cooperation between the Bureaus responsible for the construction of a ship, and also an example of the need for a change in organization and procedures which would preclude the possibility of a recurrence of a condition of this kind. After this preliminary investigation my first move was to have a study made of the steps it would be necessary to take to reestablish in this ship, and in others following the same design, the full reserve stability necessary for these ships to meet all conditions of peace and war. This of course had to be done, if possible, without impairing in any way the military characteristics of these ships in regard to fire power, cruising radius or speed. This was accomplished by a redistribution and lowering of some of the weights, the use of lighter material at high points, and the installation of lead ballast.

"My next move was to fix responsibility for this situation. Although the Bureau of Engineering and the Bureau of Ordnance must assume some contributory responsibility, the principal responsibility by regulations was vested in the Bureau of Construction & Repair. This Bureau, in assuming this responsibility, should have developed procedures which would have made it impossible for the other two Bureaus to add weight or alter centers of gravity subsequent to the time of the submission of their first estimates. To preclude the possibility of a recurrence of this situation, and to make

[35] The Bureau of Construction & Repair had also exceeded its design weights.

[36] It was a much more "glaring example" of the Bureau of Construction & Repair failing in discharging its duty as the custodian of weights, one of its principal reasons for existence.

the necessary organization changes and refinements, I took the following administrative action:

"(1) I removed the Chief of the Bureau of Construction & Repair and transferred him to other duty.

"(2) I issued an order on 11 August 1939, effecting a consolidation of the Design Divisions within the Bureaus of Engineering and Construction & Repair.

"(3) I directed that the Bureaus submit to the Acting Secretary of the Navy, for his approval, an administrative plan for effecting this consolidation not later than 1 September 1939. This was done."

This quoted memorandum should be considered along with Appendix H which is a copy of a letter signed by Secretary Edison and addressed to the Honorable James G. Scrugham, Chairman of the Naval Subcommittee of the Appropriations Committee of the House of Representatives, and dated 13 February 1940, scarcely two months later.

The amalgamation of the Bureau of Engineering and the Bureau of Construction & Repair was the work of opportunists. Secretary Edison believed that better collaboration between the design divisions of these bureaus could be obtained by combining them. Accordingly, on 11 August 1939, he issued an order for their merger and directed the heads of the two bureaus to submit plans for such a merger. These plans were submitted to a Board for review, and this Board met on 31 August 1939. The Board consisted of Rear Admiral (later Admiral) S. M. Robinson, Captain (later Rear Admiral) Paul F. Lee, engineers, and Captain Lewis B. McBride and Captain (later Vice Admiral) Edward L. Cochrane, Naval constructors. Robinson and Lee had been nominated for the Board by me to protect my interests and the interests of my Bureau. There was no indication that there was any skulduggery underway or contemplated. Everything seemed to be proceeding smoothly and as planned, the plan being that the Board was to be a smoke screen prior to the announcement of the appointment of Captain (later Rear Admiral) Joseph J. Broshek as Officer-in-Charge of this new combined design section. Broshek, at the time, was head of the Central Drafting Office of the New York Navy Yard where we were building the first high-pressure high-temperature battleships. He was a highly competent man and admirably fitted for the job. Perhaps that is why he never got it.

The opportunists took advantage of an unusual coincidence. Both Bureaus were without Chiefs. My term of office had expired in May 1939. Secretary Edison had repeatedly recommended me to the White House for reappointment but with no success.[37] Naturally there were a lot of people out looking for my scalp. The first rumor that the White House would not reappoint me was alleged to come from, as one would expect, a shipbuilder. President Franklin D. Roosevelt had removed from office the Chief of the Bureau of Construction & Repair, William G. Dubose. Robinson, McBride, and perhaps others had toyed for a long time with the idea of combining the engineers and the naval constructors. Here was an opportunity made to order. On 12 September 1939 the Board, not even having been in session two weeks, submitted its report recommending the merger of the two Bureaus and the merger of the engineers and the naval constructors, subjects never even mentioned in the precept of the Board. The Board also recommended that Rear Admiral Robinson be the first Chief of the new Bureau of Ships. Naturally I felt that my representatives on the Board had run out on me. Men have been hung for less than that. Their defection put Mr. Edison, who had supported me all along, into an isolated and peculiar position. As he said: "Even your own people went against you." The reason given me was that "the constructors would revolt" if I were the combined Chief. How I wish they could have had an opportunity! I was charged with having accomplished the downfall of Dubose and the Construction Corps, a statement which is unqualifiedly false and particularly so since it was a self-accomplished act.

37 Previously President Roosevelt had said on several occasions that he did not believe in reappointments in Washington!

On 13 September 1939 Rear Admiral Robinson was ordered as Chief of the Bureau of Engineering. On 14 September 1939 he was also ordered as Coordinator of Shipbuilding, relieving Secretary Charles Edison who had been so designated on 20 October 1938. On 5 October 1939 Rear Admiral A. H. VanKeusen (C.C.), USN, as "Chief of Construction & Repair," and Rear Admiral S. M. Robinson as "Chief of Engineering" promulgated an "Organization" order of three pages assigning "consolidated duties to divisions and officers."

So the amalgamation was effected; the greatest navy in the world no longer had an Engineer-in-Chief, and the Bureau of Engineering, in spite of its splendid and unassailable record of achievement, passed out of existence.[38]

The Bureau of Engineering was an old Bureau. It was established on 5 July 1862 as the Bureau of Steam Engineering. In it were absorbed most of the electrical functions of the Bureau of Equipment when that Bureau was abolished, and it became the Bureau of Engineering on 4 June 1920 in recognition of the fact that engineering was a rising and expanding profession. Generations of naval engineers, in seeking recognition of the importance of engineering, had fought all the way down the line from the days of the old sailing masters. In 1909, when Truman H. Newberry was Secretary of the Navy, he actually put the Bureau of Engineering under the Bureau of Construction & Repair, Chief Constructor Washington L. Capps being in charge of both, but the order for this merger was declared illegal and was rescinded. That event was only a high point in an age-old attempt of the tail to wag the dog. Essentially the same thing was done in 1939. Do you think industry would have extinguished a going concern with such a brilliant record of achievement? And two years before Pearl Harbor!

Let us look at the report of this Board for a moment because its decision was responsible for a revolutionary change in the organization of the Navy Department. The Board, as

[38] See Appendix G for Achievements of the Bureau of Engineering.

constituted, did not include any member who, up to that time at least, had distinguished himself by any original contribution to the art of organization, management, or administration. As a matter of fact, at that time two of the members were very junior and inexperienced with loaded dice. The report, consisting of five pages, was full of pious platitudes, appeals for better promotion and pay for engineers and constructors, and appeals for more building space for the new outfit. Last, but not least, it recommended that four admirals be authorized to do the work of the two displaced. It never referred to the disproportionate number of constructors to engineers, nor did it even mention the Bureau of Ordnance which, of course, should have been merged also to justify such a title as Bureau of Ships. But the "Gun Club" was too strong for them.

One quotation from this erudite report is enough. Referring to the merger of the design divisions preliminary to the merger of the two Bureaus, it stated:

"We are of the opinion that neither of the administrative plans prepared by the Chiefs of the Bureaus for the consolidation of the Design Divisions is entirely suitable for the purpose in view. We have attempted ourselves to evolve a more suitable plan but the thoroughly detailed knowledge of the ramifications of the existing organizations necessary for this purpose cannot be developed within the time and facilities available to us. Further we are of the opinion that it is disadvantageous to detail this organization too rigidly at this time. We believe that *the success or failure of such a consolidation* of Design Divisions will be largely *dependent upon the prestige, experience and above all on the personality of the officer selected* for this duty and on the confidence which he can command among the officers of the two categories [engineers and naval constructors] which are involved in the change."[39]

So there you have it. Personality plus instead of merit, knowledge, record for originality and achievement. Yes, this

[39] Italics added.

happened in the U.S. Navy Department not in some girls' school for the selection of a beauty queen. Harmonious administration had been confused with successful preparation for war.[40]

In view of this enlightened attitude of the Board, is it surprising that the Bureau of Ships underwent several reorganizations in spite of the emphasis on "personality," one of which involved the removal of the Chief of Bureau and his Deputy?

I never knew that the Board was considering a merger, nor was I ever asked to give any opinions or ideas I might have had on the subject even though I was in charge of the Bureau of Engineering. My first information about it came via the scuttlebutt. Representative James G. Scrugham (Democrat), Chairman, and Representative William Ditter (Senior Republican), a member of the Subcommittee for Naval Appropriations, begged me to approve their idea of starting a Congressional investigation, but I thought it better to do my fighting inside the Navy. Captain (later Fleet Admiral) Chester W. Nimitz, the Assistant Chief of the Bureau of Naval Personnel, made the friendly suggestion that I request a trial by General Court-Martial to establish the responsibility for the overweight destroyers.

The merger and its attendant mess did considerable damage to the morale of the engineers. When I took charge of the Bureau of Engineering in 1935 it actually ranked "with but not after" the Bureau of Construction & Repair. It had been an east coast outfit, dependent upon eastern shipbuilders. I made it a modern engineering clearing house for the Navy, dependent upon Detroit, Pittsburgh, Schenectady, and Princeton. Its engineers roamed the industrial research laboratories of the country looking for new ideas. Anyone in industry who had a new idea that might be incorporated in National defense hot-footed it, straightway, to the Bureau of Engineering. Such outstanding success breeds envy and

[40] Apologies for this apt expression to Elting E. Morison, author of *Admiral Sims,* Houghton Mifflin, 1942.

jealousy, and the Bureau of Engineering and the original EDO's had to go. My ignominious fate was a warning to pioneers, but even the most ungrateful audience I had ever worked for couldn't move back the hands of the clock of progress.

Perhaps it has not been pleasant for the reader to wade through the details of acrimonious arguments concerning high-pressure high-temperature, overweight destroyers, the merger mess, and the new American engineering ultimately installed in all types of combatant naval vessels. Nor was it pleasant to live through, although there is great satisfaction in fighting for and achieving something in which you believe. We came perilously near making a fatal mistake while we were rebuilding the Navy for World War II. Fortunately the mistake was avoided.

Great credit is due Lynn H. Korndorf, President of Federal Shipbuilding and Drydock Company, and William S. Newell, President of Bath Iron Works Corporation, for sticking with what at times looked like a sinking ship.

The beginnings of high-pressure high-temperature were started when my predecessor, Admiral S. M. Robinson, initiated the *Mahan* class of destroyers in 1933. The *Mahan* class were not completed and in operation until I had relieved him as Engineer-in-Chief of the Bureau of Engineering in May 1935. It was my lot to fight to defend what he had done as well as everything I did.

I standardized on 600 p.s.i. and 850°F, extending it to light cruisers and battleships. It is unfortunate that the term "high-pressure high-temperature" obscures the fact that what actually took place was a complete redesigning of all the machinery that goes into a naval vessel. As a result, all of our designs were standardized and tried out at sea long before Pearl Harbor. I do not know of any Bureau that was better prepared for World War II than the Bureau of Engineering.

The shipbuilders never could have supplied the quantity of machinery required by our shipbuilding effort in World

War II without General Electric, Westinghouse, and Allis-Chalmers. I brought the Foster-Wheeler Corporation into the supply field because Babcock and Wilcox were our only purveyors of boilers, and I wanted to broaden the base of supply for our war building program.

The same reason applied to my bringing in National Battery Company so that, when the time came, they could supplement the efforts of "Exide" in producing storage batteries for submarines. The submarine story is told in a separate chapter.

Before I left the Bureau, I had proved on a reengined and partly reboilered destroyer, the *Dahlgren*, that 1200 p.s.i. and 950°F was entirely feasible and urgently required. I do not know why the Bureau of Ships was so slow in following up this additional development, waiting as it did until after the war was over—and then some.

After I left the Bureau, all my advanced engineering design policies were followed by my successors in the Navy[41] and by the shipbuilding industry presumably at a great profit to themselves. The Maritime Commission also followed them in rebuilding the Merchant Marine. They have been copied all over the world, and such designs are now commonplace.

What were the origins of this revolution in shipbuilding and marine engineering? The idea to install in our naval vessels the most modern machinery American industry could produce, and to relegate the ship and engine builder to the category of an assembly yard, was brought to the Navy Department by William Francis Gibbs, of the ship design firm of Gibbs and Cox. The background of Gibbs with the Grace Line has been noted already. In other words, this revolution in the first instance was caused, not by a university, an engineering school or an institute of technology, not by a ship and engine builder, a turbine manufacturer, or even by the Bureau of Engineering, but by an individual who was a ship designer, Gibbs.

[41] See Appendix I in which the Bureau of Ships, 13 June 1943, trumpets to the world the success of high-pressure high-temperature and other engineering achievements.

Two other people were necessary in order to garner the fruits of this revolution, Secretary Edison and myself. Edison had either to support me or fire me, and he supported me with the greatest enthusiasm. If he had not supported me but had listened to the Board of Inspection and Survey, the General Board, and the shipbuilders, to say nothing of certain members of the public press, it is hard to say how much longer World War II would have lasted.

As for me, personally, I won a Pyrrhic victory. I was never invited to participate again in naval engineering. I was banished to the Naval Research Laboratory. My experiences there are another chapter.

It is the irony of fate that the new superliner, the *United States*, for which Gibbs and Cox was the design agent, should have been built at the plant of the Newport News Shipbuilding and Drydock Company. This new superliner embodies all the latest machinery which was developed as a result of the high-pressure high-temperature fight. It captured the "blue ribbon" of the Atlantic with ease. The *America* was obsolete before she left the drawing boards, and the *Queen Mary* and the *Queen Elizabeth* were obsolete when they were launched. I have stated previously that one of the private shipyards would not take a contract for one of the first of the high-pressure high-temperature battleships and, in order to get that yard to build one, I had to buy all the high-pressure machinery, including the engines and boilers, because that yard was afraid to run any risk with such dangerous machinery. That shipyard was, ironically, Newport News Shipbuilding and Drydock Company, the builder of the new Queen of the Seas, undoubtedly supplied with plenty of high-pressure high-temperature.

3

Diesel Engines

WHEN I joined the Bureau of Engineering in 1931 as Assistant Chief of the Bureau, submarines were having serious difficulties with crankshaft failures, particularly in the "S" boats. The cause was new to engineers who suddenly discovered that a propeller shaft acted as an elastic spring in responding to a variable engine torque. This torsional vibration of shafting led to crystallization and failure of crankshafts. It was the same torsional vibration I have already mentioned in connection with the vibrating whistle in the smokestack of the old armored cruiser *Tennessee.*

The most important thing in the engine room of the submarine of that day was an enormous brass plate, in full view of the throttleman on watch, on which was engraved the speed ranges or rpm (revolutions per minute) at which the engine should not be operated. These speed ranges were the critical speeds, a term which terrorized many an operator.

High-pressure air was still being used for fuel injection. The injection air compressors, which absorbed about seven per cent of the engine horsepower, worried everyone with their high maintenance requirements. Breakage of valves and excessive wear were common complaints. Explosions caused by the accumulation of oil vapor in the inter-coolers were by no means uncommon. They were rightfully treated like a sick baby with the whooping cough, and any steps toward their elimination were greeted with approval. Solid injection of fuel was still in the talk stage, with the British "common rail" system decidedly in the doghouse on account of its smoking the exhaust.

My predecessor as Chief of the Bureau of Engineering, Admiral S. M. Robinson, had obtained an appropriation of $500,000 to invest in the development program of new diesel engines. These were to be of higher speed, with many cylinders, with solid injection of fuel to each cylinder individually, and suitable for mass production.

Following the lead of European builders, our submarine engines were always direct coupled to the propeller shaft, and therefore their speed did not exceed about 400 rpm. The new program demanded not less than 700 rpm. Five engine builders responded to the Navy's invitation for bids and, of the five new engines built, only one survives to date in its descendants. This one was built by the Winton Engine Corporation, which since has become the Cleveland Diesel Engine Division of the General Motors Corporation. This splendid organization owes its existence to the indomitable pioneer spirit of George W. Codrington, its General Manager, now a Vice President of General Motors. The engine Winton submitted for test was a 12-cylinder "V" type, Model 201, 2-cycle engine of 950 horsepower at 720 revolutions per minute.[1] Its features included exhaust valves in the head, uniflow scavenging, unit type of injector, combination fuel pump and injector, and a welded steel housing.

After subjecting this experimental engine to exhaustive tests at the U.S. Naval Engineering Experiment Station at Annapolis, the engineers at the Bureau, including myself, recommended it for further development, and the Chief decided to go along with it in spite of the unfavorable report submitted by the Naval Experiment Station. For once it may be recorded by historians that a Bureau was right.

The first Winton diesel installed in our submarines, however, was a V-16 cylinder, Model 201-A. This first submarine engine was designed at the same time General Motors was designing a straight-eight engine for the "Zephyr" (Chicago, Burlington and Quincy Railroad). All of these

[1] *History and Development of the 567 Series General Motors Locomotive Engine,* by E. W. Kettering, presented before A.S.M.E., November 29, 1951.

engines, of course, were built from preliminary designs made at the General Motors Research Laboratory (Charles F. Kettering). Furthermore, this General Motors Research 2-cycle, 8 by 10 engine marked the establishment by General Motors of its basic 2-cycle engine design and has been successfully used in the full line of diesel engines built by General Motors since. The brilliant subsequent service of the Model 201-A in the Navy fully vindicated its adoption.

Two other engine builders who originally refused the Bureau's invitation to participate in the new program later submitted suitable designs of their own—the Hooven, Owens, Rentschler Company (H.O.R.) built a two-cycle double-acting 1300 horsepower, 700 rpm engine of the German M.A.N. design; and the Fairbanks Morse Company designed and built an original opposed piston engine of 1300 horsepower and 700 rpm. This was well before Fairbanks Morse entered the diesel locomotive field. I still remember my original apprehension at the use of silent chain to transmit the torque of the upper shaft to the lower crankshaft coupled to the load. It worked, and was replaced later by gearing for economic reasons.

Thus we had three engines suitable for submarine drive, except that their higher speed demanded a change from the conventional direct coupling to the propeller shaft. The idea of electric drive for submarines had long been toyed with in the Bureau—now its opportunity had arrived.

The new submarine construction program (1933), based on the availability of the new diesel engines, was used to invite bids on an electric drive. Cleveland Diesel Engine Division of General Motors Corporation was the successful bidder and received the contract for the electric drive equipment of the four pioneer submarines, *Shark, Tarpon, Porpoise*, and *Pike*, with the Elliott Company as subcontractor for the generators, motors, and controls.

For the next year's program we had a choice of engines and electrical equipment. The *Plunger* and *Pollack* received Fairbanks Morse opposed piston engines and Elliott electrical equipment; the *Perch* and *Pickerel*, another edition of the Winton, or General Motors, engines with General Electric motors and generators; and the *Pompano* was equipped with H.O.R. double-acting engines and Allis-Chalmers electrical gear.

Throughout the wartime submarine-building programs, these three engine builders and the three named electrical manufacturers supplied the Navy with the necessary propelling plants.

There was, however, in the next two years, an attempt to deviate from the straight electric drive with four generators and four or eight electric motors geared to the two propeller shafts. One engine per shaft was geared direct to the propeller shaft. This coincided with my assumption of the duties of the Chief of the Bureau, and well do I remember that I had hardly relieved my predecessor when I received a personal letter from him saying that he had changed his mind about the "composite drive," as it was designated then, and now he felt it was a bad bet. And here I was—stuck with a contract.

In "composite drive," two of the four main engines are coupled directly to each of two propeller shafts and the other two engines are connected to the propeller shafts by electric drive.

I promptly had a conference with George W. Codrington and his torsional vibration expert to explore the possibility of replacing the new drive if it became necessary. As a result of this conference, I decided to stick with it. As a matter of fact, it worked very nicely and made some staunch friends among the operating men in the submarines. Nevertheless, electric drive was installed in subsequent classes of submarines because the majority of submariners favored it. Anyhow, what is one failure—if one chooses to regard it as such—when you are trying to redesign all the machinery of an entire Navy in time to beat the enemy to the punch in the world war that was coming.

All was well with electric drive in submarines until they got to sea; then the commutator risers started breaking off as fast as they could be replaced. Fortunately the General Electric Company came to the rescue by lending us an engineer who was an authority on commutation, and he solved our problem. Nevertheless, lasting credit is due to the Elliott Company for having pioneered in the realm of electric drive for submarines.

This difficulty with the electric drive was a ripple on a mill pond compared with a row which later broke loose concerning the new submarine engines.

Those who didn't accuse me of trying to wreck the Navy with high-pressure high-temperature steam claimed I was bound to do it by dieselizing it. As a matter of fact, the only diesel contract I ever regretted was the case where two larger engines were not ready for our first new submarine tender, and eight submarine engines, driving generators, were installed instead. This, however, gave us a chance to test direct-current electric drive against alternating current in sister ships. But even these delayed larger engines, later installed in the oiler, *Maumee*, gave an excellent account of themselves in war service.

The Secretary of the Navy, Mr. Charles Edison, was deluged with reports against me and against the new submarine engines from the Commander-in-Chief of the Pacific Fleet all the way down to an "expert" on submarines in his own office. An expert seems to be a man who knows all about what has happened but doesn't know what is going to happen. The Commander-in-Chief complained that his submarines with "Bowen" engines were broken down from Alaska to Panama, so I contracted for more submarines just like them. It is remarkable how quick we are to complain about the few in trouble and disregard the many giving excellent service. I was sure I was on the right track and had more confidence in the design and manufacturing facilities of General Motors and Fairbanks Morse than I did in the engineering opinions of the Commander-in-Chief of the Pacific or Mr. Edison's "expert."

After going over the reports from the submarine engineers, at least I had the satisfaction of knowing that no one was sure which was the worst engine. That was real progress. In fairness to all, however, I must say that we did declare a year's holiday on the construction of the double-acting H.O.R. engine until George A. Rentschler, its dynamic sponsor, had time to trace the trouble to its source and correct it.

I had troubles of my own in getting information about the General Motors engines. As soon as a submarine skipper having these engines got into port, he would call up George Codrington and tell him what he needed to keep operating. Spare parts would be shipped promptly or a service engineer flown to the site at no charge. With a service like that, the skipper felt that he would be wasting his time and mine to report his engine problems to me.

At one time we had a lot of piston ring trouble. Since, in the meanwhile, General Motors engines similar to our submarine engines were running successfully in locomotives out of Chicago, my young men haunted the round houses and rode diesel locomotives all over the country to get the latest "dope" on piston rings. The cylinders of Fairbanks Morse engines developed hot spots and cracks in the combustion space after some considerable service. But I stuck with them nevertheless. There is such a thing as being an impractical perfectionist. Not much would be done in this world if we had to wait for the perfect machine to do it.

One very important thing the critics of the new submarine plant overlooked in evaluating its merits for wartime service was that, in a desperate situation thousands of miles away from base, one engine could be cannibalized to keep the other three running. Interchangeability, of course, is the keystone of mass production. Furthermore, the component parts of the new higher speed engines were much lighter and could be handled with ease by one or two men. Of great importance to the operating men was the fact that the ominous and menacing brass plate with its criticals disappeared from the engine room.

Our new submarines played a vital role in the war with Japan and Germany; they virtually choked Japan to death unaided. In addition the excellent performance of their engines gave tremendous support to the dieselization of our railroads, an accomplishment which pulled many a railroad out of the red and into the vigor of profitable operation. And in the old engines, burning coal without extracting its useful chemicals was a waste of natural resources. The diesel locomotive put an end to that, and our coal reserves will last just that much longer. Another example of one of the many different ways the Navy serves our country.

Another important change was the dieselization of Navy motor boats. Formerly they were driven by gasoline engines, with explosions and fires continuously causing much damage and occasional loss of life. Gasoline piping is very difficult to keep tight, and leakages into the bilges were inevitable. But the greatest danger accompanied fueling of the boats when gasoline vapor filled the engine enclosure.

We learned early that the metal nozzle of the filling hose had to be grounded to prevent an electric spark from causing an explosion. We installed fans and required that, before an engine was started, the engine room had to be thoroughly ventilated. It was required practice in the fleet for every ship's boat in turn to come alongside the gangway every morning after fueling and the Officer-of-the-Deck would detail an assistant to inspect the bilges. Smoking was prohibited in the motor boats as a matter of course, but human nature being what it is the explosions continued—and they were not all caused by smoking. As Fleet Engineer of the Battle Fleet I used to say that the only safe way to inspect a gasoline boat was to have the Officer-of-the-Deck throw lighted cigarette butts into the bilges to see whether or not the bilges were being maintained dry and free of gasoline.

After I became Chief of the Bureau I decided to replace the gasoline engines in motor boats with diesel engines. The Chairman of the Navy Appropriations Section of the House Appropriations Committee, Representative, later Senator, James G. Scrugham of Nevada, an engineer by education, was heartily in accord with my intentions. He was a great prospector and, after Lake Mead was formed by the Boulder Dam, he used to cruise continually in a gasoline boat around that lake prospecting for minerals on its shores. One day his boat blew up; Scrugham suffered serious burns and had to swim ashore. It is doubtful if the Navy could have dieselized its boats as rapidly if it had not been for that explosion on Lake Mead.

Before the summer of 1933 the Navy manufactured at the Norfolk Navy Yard three different sizes of gasoline motor boat engines from designs by Van Blerk, a well-known marine engine designer. In seeking diesel designs, I made it a condition of acceptance that a manufacturer must submit designs for all three sizes of engines and the three designs must be similar. The purpose was to reduce the number of parts as much as possible. The Buda Company was a successful manufacturer of high-speed engines, and it was selected to provide the Navy with diesel engines for motor boats and also the necessary drawings to permit these engines to be manufactured at the Norfolk Navy Yard. At the same time the Navy took out a license under Lanova patents for a pre-combustion chamber which produced proper atomization and penetration of oil charge as well as reliable ignition, and we insisted that it be incorporated in the new engines. Many thousands of these diesel engines were built during World War II, both at Norfolk Navy Yard and by the Buda Company.

Today gasoline explosions and fires are restricted to pleasure craft and still make headlines in the newspapers.

The last thing I did about diesel engines before my term as Engineer-in-Chief of the Navy came to its end was to buy "Boss" Kettering's latest creation, the "pancake engine." Charles F. Kettering, Vice President of General Motors, was in charge of their Research Laboratories in Detroit where many a dream became a reality due to the persistence, enthusiasm, and vision of this remarkable man. The perennial demand for reduced engine weight challenged Kettering's fertile imagination, and he conceived a vertical-shaft engine consisting of four radial engines of four cylinders each, stacked like pancakes. The nickname "pancake" has stuck to it, even to this day, and stands for accessibility and ease of maintenance of each individual cylinder.

George W. Codrington agreed with Kettering and was anxious to develop the proposed engine, but Mr. R. K. Evans, then Vice President in charge of General Motors' diesel engine activities, would not go along because in his opinion the engine had no commercial future. Well do I remember a stormy meeting one afternoon at the General Motors Laboratories with Evans, Kettering, and Codrington when I steadfastly maintained that no one could look over the horizon and predict what would happen to an engineering development. Didn't the expert tailors laugh and persecute Isaac Singer for his sewing machine? Then we argued at length on the probable cost of development of such a new engine. I offered to assume, for the Government, half of this cost, whatever it might be. But Mr. Evans would not accept my proposition and only reluctantly agreed to undertake the development with the Government's share of the expense to be $250,000. The final cost turned out to be well over a million and a half, with the Government paying only $250,000. This was the best afternoon's work I ever did for Uncle Sam.

The pancake engine was widely used in World War II in light submarine chasers. It weighed only four and a half pounds per horsepower and, despite such light weight, gave an excellent account of itself in war service. Furthermore, as late as June 1951, the newspapers carried an account of the Secretary of the Navy Matthews bragging about the most wonderful diesel ever installed in our submarines, weighing only half as much per horsepower as did the ones we had used in the last war.[2] That engine is the lineal descendant of the pancake engine. Perhaps I should add with pardonable

[2] The pre-General Motors diesel submarine engine, generally of German derivation, weighed about 100 pounds per shaft horsepower, the multi-cylindered engines (General Motors and Fairbanks Morse) about 18½ pounds, the "pancake," about 4½ pounds, and the radial (pancake) diesel for submarines, 9.3 pounds.

pride that this engine is now being manufactured extensively for commercial use.

Is it any wonder that, having offended so many sensibilities by my activities on behalf of high-pressure high-temperature steam, new high-speed lightweight diesel engines for submarines, the thoroughly unconventional pancake engine, not to mention radar, I received no offers from the Navy to continue in engineering when my appointment as Engineer-in-Chief of the Navy and Chief of the Bureau of Engineering expired.

I was fortunate indeed, and doubly fortunate was the Navy, when Mr. James Forrestal became the Secretary of the Navy. My experiences with him constitute the chapters on plant seizures and the Office of Naval Research.

I must not close this chapter, however, without acknowledging what a pleasure and profit it was to work with two such outstanding diesel engineers and submariners as the then Commanders W. Durward Leggett and Lisle Small.

4

Naval Research Laboratory

COINCIDENT with my orders to the Naval Research Laboratory in the fall of 1939 the Bureau of Engineering was being merged with the Bureau of Construction and Repair to form the Bureau of Ships (June 1940), but the Laboratory which had been under the jurisdiction of the Bureau of Engineering was placed under the Secretary. My orders made me Director of the Naval Research Laboratory and Technical Aide to the Secretary of the Navy. The Technical Aide designation was made to save face for me, but it was made secondary in order not to give it too much prominence.

Such a storm was raised because of my successful attempt to modernize machinery in our naval vessels in line with the latest developments of the industrial laboratories of the great corporations that I naturally became, in 1939, not only an extremely unpopular person but a political liability as well, and everyone knows what happens to a political liability in Washington!

In the words of one of our most disreputable commentators, I was "banished" to the Naval Research Laboratory as Director where, according to the commentator, I would be so far away and would operate in such a medium that I could not possibly be of any further damage to the Navy. What an interesting comment on the state of appreciation of research at that time—that one who had wrecked the Navy should be placed in charge of research.

My detail to the Naval Research Laboratory of course ended my engineering career. It was a high price to pay for principles, but at least I had the satisfaction of knowing that my country had not invested billions of dollars in an obsolete Navy which is what would have happened if the Navy had not adopted high-pressure high-temperature steam.

As a matter of fact, as Engineer-in-Chief I had kept close track of the work going on at the Naval Research Laboratory, since it was under the Bureau of Engineering, particularly in the fields of radar, sonar, and nuclear physics, and this appointment gave me an opportunity to go still deeper into those projects. Radar means *ra*dio-*d*etection-*a*nd-*r*anging. The term was coined by Lieutenant Commanders (now Captains) Frederick R. Furth, who was then in the office of the Chief of Naval Operations, and Samuel M. Tucker, while he was on duty in the radio section of the Bureau of Engineering.

Radio and Radar—1922 to 1940

In 1922 Dr. A. Hoyt Taylor and Leo C. Young of the Naval Research Laboratory noticed, in receiving radio signals transmitted across the Potomac River at five meters wave length, a distortion of the signals when a vessel was approaching or leaving the vicinity of the transmitter or receiver. Actually the vessel was approaching from down stream, approximately opposite the present site of the Naval Research Laboratory, with the receiver located at Hains Point and the transmitter at the Anacostia Air Station.

During the 1920's, immediately following the establishment of the Naval Research Laboratory, Taylor and Young continued studying wave propagation, carrying out extensive and pioneer work in this field—especially in the higher frequencies. It was during this period that the skip distance theory was evolved, checked, and many Heaviside layer measurements made.

From this work, with the invaluable help of Dr. Edward O. Hulburt, some of the basic theories were evolved relative to high-frequency radio propagation. This early work was all carried out with relatively low-angle radiation involving various distant points. In 1925, from theoretical considerations after reviewing this type of work, Gregory Breit and

Merle A. Tuve of the Carnegie Institution thought it would be possible to get signals back vertically from the Heaviside layer, provided the frequency was below the critical frequency. The wave lengths employed were 37½ and 75 meters. The Laboratory concurred and cooperated in performing the experimental work. In this connection, L. A. Gebhard built a modulator and revised one of our high-power high-frequency transmitters to provide sufficiently short signals for the experiment while Young built a suitable receiver. Thus the equipment was built by the Naval Research Laboratory and the original experiment carried out there,[1] demonstrating that signals could be reflected and received back vertically from the extensive ionized sheet known as the Kennelly-Heaviside layer. Shortly after the experimental work had been underway, M. H. Schrenk of the Naval Research Laboratory was able to shorten and sharpen the pulses considerably by making use of the multi-vibrator principle. The pulses so produced were still very long as compared to those required for successful radar.

On 24 June 1930 Leo C. Young and L. A. Hyland of the Naval Research Laboratory discovered that an airplane could cause interference with reception when it passed through electromagnetic radiation. The possibilities of the radio detection of planes was recognized by Taylor, Young and Hyland then and there.

By 1931 the Bureau of Engineering directed the Naval Research Laboratory to "investigate the use of radio to detect the presence of enemy vessels and aircraft. Special emphasis is placed on the confidential nature of this problem."

In the year 1933 the phenomenon of reflection of radio frequency energy by ships and aircraft was familiar to many

[1] In regard to the early experiments to determine the height of the Kennelly-Heaviside layer, the transmitting and receiving apparatus for these measurements was actually developed and tested at the Naval Research Laboratory. However, since such measurements at that time required physical separation of the transmitter and receiver, the transmitter was located at the Naval Research Laboratory and the receiver at the Magnetic Observatory of the Department of Terrestrial Magnetism, Carnegie Institution, 5241 Broad Branch Road, N.W., Washington, D.C.

observers. The technique of measuring the heights of the ionospheric layers by radio frequency pulses of the order of .01 seconds in length was a well established procedure. The use of a similar technique by an aircraft for measurement of its height above the ground had been suggested, but no attempt had been made to apply it.

At that time a small group at the Naval Research Laboratory had been attempting to find a practical way to detect and locate ships and aircraft by radio frequency energy. This group had observed reflections from buildings, ships, and aircraft. They had developed and operated ionosphere probing apparatus. They had proposed several detection systems based on CW (continuous wave) transmission which were not accepted by the Navy. From time to time the project was temporarily abandoned as hopeless, at one time the proposal having been made officially that the work be transferred to the Army as applicable only to large area protection. In 1932 Taylor worked out a complete system of anti-aircraft defense, using detection equipment then available, and turned it over to the Army.

The reason for this belief that the future of radar would be with the Army and not the Navy is illuminating. At this particular time it was believed necessary that the transmitter and receiver must be separated from each other by a distance of not less than a quarter of a mile. The complete disclosure of the radar discoveries by the Naval Research Laboratory to the Army early in 1932 was directly responsible for the large-scale Army developments of radar subsequently.

That the effort was not discontinued permanently by the Laboratory is due largely to L. C. Young. He had discussed the possibilities of using the ionosphere pulse probe for aircraft detection, but the idea was not pursued at the time because there appeared to be no way to generate pulses sufficiently short, and, if produced, the energy from so short a pulse reflected by an object so small as an aircraft was considered to be too minute for detection at any practical range. With all this in mind Young was studying the nature and origin of key clicks in an effort to eliminate them from radio communication. With a transmitter keyed in synchronism with a cathode ray tube sweep, a cathode ray presentation of the transmitted dot envelope was obtained. With this presentation, keying transients were observed with amplitudes greater than the CW amplitude of the transmitter. This amplitude was observed even when the CW radiation was over 20 KW. Phenomenal as was the power level of these short pulses, even more startling was the short duration, this appearing to be a very small fraction of a millisecond. Mr. Young then started putting twos together and getting fours, reasoning thus:

> Using CW transmission and observing variations in the interference field due to reflection from an aircraft in flight, detection ranges up to 40 miles had been obtained with a 500-watt transmitter.
>
> Very short pulses at a power level on the order of 30 to 50 kw were observed to be radiated from a high-frequency transmitter.
>
> Therefore, it should be possible to generate and radiate pulses sufficiently *short* and sufficiently *powerful* to get detectable echo signals and accurate range measurements from aircraft at useful ranges.

Armed with this argument, Young succeeded in getting the general project continued. Having been convinced by Young of the possibilities of the pulse system, Dr. A. Hoyt Taylor vigorously backed the project, although he realized at the start that it was to be a long hard struggle. He successfully maintained the support of the Director, the Bureau of Engineering, and other departments of the Navy over a period of years.

With a decision reached on *what* to attempt, the basic concept having been worked out, there remained the problem of *how* to accomplish it. Taylor and Young assigned this problem to R. M. Page, who started to work on its solution on 14 March 1934. There is no record establishing the exact date of Young's suggestions or of the decision to attempt the pulse method, but since, of necessity, it preceded Page's assignment to the project, at least it can be said that both took place sometime prior to 14 March 1934. There is little doubt that the early solution of this problem was due in large part to the brilliant work of R. M. Page.

During the year 1934 a pulse radar system was built in the Laboratory. It was first put on the air and tested as a system in December 1934. It received echoes from airplanes at short ranges, but the band width of the receiver was too narrow to give satisfactory resolution of signals. This system operated on 60 megacycles. It is interesting to note that D. G. Fink and L. N. Ridenour set the date for similar activity by Sir Robert Watson-Watt of England as sometime early in 1935. Both ignore the existence of the Naval Research Laboratory accomplishment, and Ridenour credits the Naval Research Laboratory only for the set installed as described later on the *New York* in 1938.[2]

During the year 1935 the mathematical solution for the rise and decay times of a multi-stage tuned receiver was accomplished and a receiver was designed and built. In the fall of 1935 R. C. Guthrie joined the project to develop and build a high power self-quenching pulse transmitter. These were combined with new synchronizing and indicating devices to make a second pulse radar system operating on approximately 29 megacycles. This system was first put into operation in April 1936. Transmitter power was approximately 7 kw, the receiver time constant approximately 5 microseconds, echo patterns were excellent and ranges of 25 miles were obtained on aircraft.

In 1935 the Naval subcommittee of the Appropriations Committee of the House of Representatives, of its own initiative, added to the appropriation for the Naval Research Laboratory the sum of $100,000 for further research in radar because of its intense interest in radar development.

[2] D. G. Fink, *Radar Engineering*, McGraw-Hill, N.Y., 1947, p. 7, and L. N. Ridenour, *Radar System Engineering*, McGraw-Hill, N.Y., 1947, p. 14.

The committee had repeatedly made inspections of the Naval Research Laboratory and had given special financial support to its work. During most of this period, Representative James G. Scrugham of Nevada (later Senator), a former engineer, was Chairman of the committee. Other strong supporters on this committee were Mr. Charles A. Plumley of Vermont, Joseph E. Casey of Massachusetts, Harry R. Sheppard of California, and William Ditter of Pennsylvania. While not a member of any Naval committee, Senator Owen Brewster of Maine was most interested and helpful always, even going to sea on the trials of a destroyer in order to learn about high-pressure high-temperature.

In May 1936 A. A. Varela joined the project and started the development of a 200 megacycle radar system. This system was first put into operation, with receiver and transmitter working into a common antenna, in August 1936. Its operation was completely successful.

In June 1936 representatives of the Bureau of Engineering witnessed a demonstration of aircraft detection equipment at the Naval Research Laboratory and, as Chief of the Bureau, I directed that plans be made for the installation of a complete set of radar equipment, as it then existed, aboard ship.

On 17 February 1937 Assistant Secretary of the Navy (later Secretary) Charles Edison and Admiral William D. Leahy, USN, Chief of Naval Operations (later Chief of Staff to the Commander-in-Chief of the Army and Navy), witnessed at the Naval Research Laboratory a demonstration of the detection of aircraft by one of the early 200 megacycle radar sets.

The 200 megacycle work was followed up vigorously by the Page-Guthrie-Varela team, resulting in a series of tests aboard the *Leary* in April 1937, and design and construction of the first fully engineered Naval service radar equipment in 1938, the now famous XAF, a high power search radar set using the pulse principle. This set which is still in existence and operable was installed in the battleship *New*

York in December 1938.[3] Reports from its fleet test were awaited anxiously. In passing, it should be noted for the record that a Doppler Radar, using a magnetron and working in the neighborhood of 10 centimeters, was taken along on the April 1937 *Leary* cruise with which signals from shipping and sea return were obtained.

This first radar set had a large rotating antenna approximately 17 feet square. It was known as a "flying mattress" and was not very welcome among those who placed great moment on the appearance of their ships.

About this time the Bureau of Engineering gave a project to the Radio Corporation of America to build an experimental radar, operating near the 400 megacycle band, with the hopes that it could be produced in time to take the cruise on the *New York* alongside the Naval Research Laboratory set. Of course, necessary disclosures were made to the Radio Corporation of America to assist them in this work.

It was quite a task for the Naval Research Laboratory to have a sea-going radar set ready in December 1938 and, probably, an even greater task for the Radio Corporation of America. Nevertheless, both sets were ready for installation by the time specified. The Laboratory set performed very much better than the RCA set as might be expected on account of greater familiarity of the Laboratory with the subject. The Laboratory set operated nearly twenty hours a day for almost three months. It was difficult to keep the 400 megacycle equipment operating long enough to get good information. This of course naturally followed from the hastiness with which this set had been prepared.

Perhaps the first tactical demonstration of the possibilities of radar took place in 1939 under forces commanded by Vice Admiral A. W. Johnson. Johnson had this opportunity not only because of his interest in the new art but also because he did not object to his flagship displaying the unsightly "flying mattress." In fleet exercises, during a destroyer attack on the *New York*, one searchlight was trained

[3] From "Notes on Early Radar History," by Dr. R. M. Page.

by XAF radar information before it was turned on and, when it was turned on, it illuminated the lead destroyer of the attacking squadron at a distance of 9,000 yards. A second demonstration convinced all concerned that radar could detect and warn of night attack by destroyers. Admiral Johnson was so impressed with the radar detection of the destroyers that he reported to the Navy Department that radar "is one of the most important military developments since the advent of radio itself."

Subsequent to the full scale sea-going tests of this XAF on the *New York* in 1939, this set became the production prototype for the radar model CXAM.

Meanwhile, arrangements were being made to purchase six models of the Naval Research Laboratory radar equipment on what was the first contract the Navy let for radar. The records of the Laboratory show that complete disclosures of the details of this equipment were made on 18 May 1939 to representatives of the Bell Laboratories and on 19 May to representatives of the Radio Corporation of America. Through competitive bidding, the contract for the construction of six sets was awarded to the Radio Corporation of America in 1939, the Naval Research Laboratory to give all possible help. The Laboratory delivered to the Radio Corporation of America the original XAF to be copied by it in the manufacture of the CXAM. Subsequently twenty CXAM equipments were built by RCA as "Chinese copies" of the XAF. Nineteen of these equipments were installed and in service in the fleet at the outbreak of war with Japan. It was on these sets that the Navy relied with such gratifying results during the first years of the war.

Among other discoveries of this time was the fact that reflections from projectiles in flight could be depicted on the radar screen. Later developments in radar still further improved this technique, a technique which proved to be very valuable during the war.

It was also discovered that keeping ships in position in formation was possible with radar, although at that time it was not satisfactory at distances under 1200 yards. This technique later became satisfactorily developed with the advent of much higher frequencies, shorter pulse lengths, and tighter beams.

It was necessary to adopt measures to prevent the interferences of radar with communications systems.

All of these things are mentioned in order to show the tremendous amount of work in connection with radar which did not directly contribute to the development of radar.

People are so accustomed to the fine discrimination which now characterizes radar equipment that they probably would have been dismayed at the lack of definition which characterized these early sets. However, even at this stage, ranges were reasonably accurate and reliable, and bearings could be obtained on objects on land and sea. Even then it was obvious that the optical range finder had not only met a competitor but that it had met something extremely more accurate, at least for long ranges. The accuracy of radar measurement is independent of distance, while the accuracy of the optical range finder decreases with distance. The measurement of range by the range finder depends on the base line which extends from the line of sight at one end of the range finder to the line of sight at the other end. As the angle subtended by this base line continually decreases with increasing range, the error obviously increases with the reduction in the subtended angle or increase in range.

Meanwhile the Naval Research Laboratory had other problems besides radar design.

In 1938 the Naval Research Laboratory was endeavoring, first, to get radar into production and into the fleet; second, to expedite the completion of radio homing devices for carrier planes and short-based planes; third, because of revived interest, to continue the development started as early as 1920 of radio controlled devices for pilotless planes or drones in order that fleet gunnery would have the opportunity to simulate the problems of actual combat with planes; fourth, to extend the use of very high frequency into communications; fifth, to develop high-frequency direction finders; sixth, to complete the work already underway on pulse altimeters and to develop instruments to measure drift; seventh, to develop anti-submarine detection devices of a portable or semi-portable nature which could be used to protect harbors or advance bases against submarine raids; eighth, to develop recognition equipment. As early as June 1938 the Laboratory successfully tested voice communication equipment operating in a 300-360 megacycle band either from plane to ground or from plane to ship with satisfactory ranges over a hundred miles, a very early use of rather high frequencies for airplane communications.

Under date of 8 December 1939, while I was Director of the Naval Research Laboratory, I informed the Acting Secretary of the Navy that new work on detection (radar) with higher frequencies would, as soon as suitable tubes were available, permit great reduction in the size of rotating antennas and I observed: "It is hoped that the Navy will shortly be in a position to use radar on planes for bombing. The 500 megacycle equipment now under development functions well at 20,000 feet. . . . Development is proceeding along two lines, one for long range detection and the other for higher frequencies for altimeter work and fire control, including bombing."

Dr. A. Hoyt Taylor, in his *Radio Reminiscences*, comments: "This letter is of interest because it anticipates quite exactly the trend of radar development during the early years of the war."

The year 1940 marks a definite turning point in the history of the development of radar and the Naval Research Laboratory through the introduction of new agencies and a greater awareness of the imminence of global warfare.

On 26 February 1940 the Naval Research Laboratory laid special stress on the development of higher frequencies and the importance of recognition equipment associated with detection equipment. It also stressed the necessity of expediting the application of radar to fire control and the development of the repeater unit, particularly the polar chart, better known as the Plan Position Indicator (PPI).

No one at this time could possibly anticipate the eventual magnitude of the radar work in which we were involved. Under date of 7 May 1940[4] the Bureau of Ships informed me that "The Bureau considers that the inherent possibilities of the radio echo methods are such as to offer compelling reasons for pressing the development of all phases of this problem (insofar as consistent with reasonable economy)." The Laboratory replied to this admonition by requesting that its funds be increased two and a half times over the amount allowed for the preceding year.

On 28 May 1940 the Bureau of Ships informed the Chief of Naval Operations that, since developments of radar would be quite rapid and, consequently, new equipment would soon become obsolete, to observe a reasonable economy would be a better long-range solution.

In the following month, I reported to the Bureau of Ships seven important phases of radio detection work of which only three were very active due to lack of personnel and facilities at the Naval Research Laboratory. These three were: (1) surface and aircraft detection work; (2) basic development; and (3) detection of aircraft by submarine.

On 1 June 1940 the Chief of Naval Operations expressed to the Bureau of Ships his satisfaction with its policy in regard to the development of radar at that time. He added, however, that "in view of the present international situation, it is desired that every effort be made to expedite the completion of development of the project in all its phases and to commence procurement at the earliest possible date that is justified by the success obtained in every subdivision of the project. Accordingly, it is suggested that expansion of facilities and personnel be undertaken as soon as funds are in hand."

[4] The merger of the Bureau of Construction and Repair and the Bureau of Engineering into the Bureau of Ships began about September 1939 with the merger of the respective design divisions, but the merger was not complete and the new name not adopted until June 1940 although, during this period, both Bureaus were under the administration of Rear Admiral (later Admiral) S. M. Robinson.

In 1940, when it appeared that the next thing to do in radar was to construct an installation for submarines, I called Dr. Taylor into my office and asked him to think about the possibilities of such an installation. My idea was that the radar would sound a general alarm signal, as a submarine does not have time after it sees a plane to do anything but dive. I told Taylor to think about it, to worry a lot at night, and when he had finished his thinking on the matter to come back and tell me the results.

In about two weeks, Taylor returned to my office and told me he thought there was a remote possibility that we could devise a radar installation for submarines, but that he would need half of our radar force of six for this work and that he could not tell inside of three months what the prospect of success would be. I assigned the three men to this work. Needless to say, eyebrows were raised in the Bureau of Ships at my taking the bull by the horns in this respect without consulting them.

Three weeks later Taylor returned again and told me he thought there was a 50-50 chance of producing radar for submarines, and in three months such a radar was produced. This was unquestionably the first radar by far to be installed in submarines anywhere. The British had not even considered it up to the time we made our mutual disclosures.

One of the interesting by-products of this installation of radar in submarines was the discovery by our own engineers that the radio periscope of the submarine did not even transmit radio communication with the highest possible efficiency so that, in designing transmission for radar up the radio periscope, we were able also to improve considerably our transmission of radio communication through the same periscope.

Another development in 1940 was the formation of the National Defense Research Committee in June. This of course ultimately became a part of the Office of Scientific Research and Development. I was the original Navy Department member of the National Defense Research Committee. The Naval Research Laboratory made complete disclosures of all work underway and contemplated to the National Defense Research Committee. I had hopes that the National Defense Research Committee would use the Naval Research Laboratory as a center for its rapidly expanding work, particularly in radar, since the Laboratory was the father of radar. However, this was not done, and the Radiation Laboratory at the Massachusetts Institute of Technology and the work at Harvard University in electronics became of such an enormous magnitude that it is probable that, mechanically, things worked out for the best. The records of the Laboratory indicate that Drs. Karl T. Compton and Alfred L. Loomis visited there in July of 1940 and, for the first time, saw Naval radar in operation.

The records also show that as early as May 1941 Dr. I. I. Rabi of the Radiation Laboratory accompanied Dr. Robert F. Bacher to become informed of the Laboratory's progress in radar, particularly with respect to the Plan Position Indicator development.

Commercial research laboratories such as Bell Telephone, Westinghouse, and General Electric had known about the work going on at the Naval Research Laboratory and the Signal Corps Laboratory at Ft. Monmouth for a great many years, but none of these great corporations visualized, nor did anyone else for that matter, to what extent radar would finally be developed and utilized. As engineers from the Bell Laboratories once said: "We are delighted to know that you are doing this work; it should go on. Our directors would never allow us to engage in this class of work which does not appear to have any commercial possibilities."

In August 1940 I had lunch with Mr. Charles E. Wilson, president of General Electric, and his government representative, Mr. Bryce Blair, in Washington. I realized that this nation was faced with limited radar production facilities in the event of war. I informed Mr. Wilson that either the General Electric Company was crazy not to get into radar or that I was crazy for advocating it and, furthermore, I wished that he would ask his scientists up in Schenectady who was crazy. He said he would be delighted to do this as he was on his way to Schenectady when he left Washington that afternoon. About two weeks later a distinguished party of twenty or more from General Electric, headed by Mr. Wilson himself, visited the Naval Research Laboratory. Other men in the party were Dr. Ernest F. W. Alexanderson, the inventor of the Alexanderson alternator for radio communication, and Dr. W. R. G. Baker, who was in charge of the General Electric Radio Division. Mr. Wilson was very much impressed and so was Alexanderson. Mr. Wilson told me that he would put the General Electric Company into the radar business because he thought it was the patriotic thing to do. I told him that not only was it the patriotic thing to do but it would enable him to participate in the beginnings of a new billion-dollar industry. At one stage of the inspection I found Dr. Alexanderson musing by himself. He told me that he had heard about our work in radar at the Laboratory, that he always thought someone would do something along these lines, but that he never dreamed it would be done exactly as we had disclosed to him.

In about another two weeks, Mr. Wilson made another visit to the Naval Research Laboratory accompanied by an even larger party. As a result of this second inspection he told Dr. Baker to spare no expense to put the General Electric Company in the same position of competence in regard to radar as the Naval Research Laboratory.

Among other things, we showed the party how radar could be transmitted from a single bar antenna resembling a steam condenser tube and the reflections picked up and recorded by the same rod. Alexanderson said that was the most wonderful thing he had ever seen in his life—to think that a brass rod about 15 feet long could faithfully record radar reflections. That of course represented the disclosure of the first radar equipment making use of a submarine periscope antenna.

Later, in 1942, I was invited to lunch with Mr. Wilson in New York City. Present at the luncheon, among others, were Mr. Gerard Swope, former president of the General Electric Company, Mr. Bryce W. Blair, General Electric's

Washington, D.C., representative, and Dr. Baker. I described some of the astounding results obtained by radar in the early battles of the Pacific. Mr. Wilson was very much impressed and announced that he had decided to originate no more new commercial radio sets but to devote all the design talent of General Electric to radar. Dr. Baker was terribly upset by this announcement. He stated that he had worked for years to get his Radio Division out of the red, that he had succeeded and that now, in no time at all, General Electric would lose all of its commercial business. Wilson replied that his mind was made up and that what he wanted to know from Baker was whether their new radar plants should be located at Bridgeport, Connecticut, Schenectady, New York, or elsewhere, and how much money he wanted for the necessary expansion in electronic facilities.

We were soon in the market for a 200 megacycle search radar of much higher power than the original sets and for installation on destroyers. Of course, for this application the antenna would have to be much smaller. A contract was let with the General Electric Company for the production of several hundred sets and these were known as SC Radar Equipments. Unfortunately, twenty or more of these sets were sent to sea before the Naval Research Laboratory was able to obtain a set for test, and we were only then able to obtain a set for test because there were some deficiencies which had arisen in operation probably due to the haste in which they had been constructed. These sets, however, were soon corrected and proved to be very valuable. They were widely used, especially in convoy work in the Atlantic, and many days in the aggregate were saved because convoys were not delayed by fog. Later on, all of these General Electric sets were replaced by newer and improved models; nevertheless, they served in a pinch in both the Atlantic and Pacific.

By the beginning of 1940 the Naval Research Laboratory had introduced into the search radar system most of the important parameters and variables that exist in radar as we know it today. The important thing to bear in mind at this point is that the implementation of these ideas was necessarily crude due to the early state of research and to the small number of people (in the light of the thousands who continued the work during the war years) who were responsible for introducing these features into the radar system.

The frequency at which a radar set operates determines the specific uses for which it is best fitted. The features which really needed improvement at this time were such things as an increase in the available power output, in receiver sensitivity, and in the resolution of the radar. This increased definition can be obtained by only two methods. One is to increase the frequency at which radar will operate; the other is to increase the size of the antenna. As has been pointed out earlier, the antenna used at 200 megacycles was already so large that fleet personnel were not too happy about its installation on the top of a mast of a ship. It was clear that to do the best possible fire control job it was necessary to increase the resolution of the antenna beam used and, due to the relationship existing between the size of the antenna, the frequency, and the angular width of the radiated beam, it was necessary to go to very high frequencies in order to get sharply directed beams with reasonably sized antennas. Also, to detect small objects on the surface of the sea and to resolve two closely spaced objects, very high frequencies with very sharp beams and short pulses were again required. It was obvious, therefore, that further improvements in radar performance were undoubtedly to be had only by the use of much higher frequencies.

The Laboratory had been aware of this requirement since nearly the beginning of the radar development but found it impossible to implement it because of the lack of a powerful but stable generator of power at frequencies on the order of 3,000 megacycles, i.e., 10 centimeters.

The extension of the use of the radio frequencies from a few hundred megacycles up to a few thousand megacycles was a truly major undertaking. Early work of this nature was underway in the 1930's at the Naval Research Laboratory as well as in other laboratories in this country and in

Europe. None of this work, however, produced an oscillator of unusual power output.

The results of the Naval Research Laboratory's early research work carried out on wave propagation, including the Kennelly-Heaviside layer studies as well as that of others, had been reported in many scientific journals of international circulation.

In September 1940 we traded information with the British via their mission headed by Sir Henry Tizard, and a member admitted to me that the above-mentioned articles resulted in their beginning work on what they called radio locators. This is a convenient time to examine the progress in radar in the United States and Great Britain.

Our experiences with the British were very interesting indeed. The British had concentrated their efforts on ground-installed search radars and on airborne radar. One of their important disclosures was an airborne search radar set operating at 175 megacycles. A considerable number of these sets were obtained and installed on large patrol planes, especially in the Aleutian Islands. In passing it may be said that radar took the Aleutian Islands from obscurity and made them into a highly strategical area.

Because of its low frequency, the British radar had a very large antenna. At the time this disclosure was made, they were using one antenna for transmitting and a second one for receiving. The method known as "duplexing" whereby one antenna is used for both sending and receiving was unknown to them. The head resistance of these two antennas installed on British planes was sufficient to cut down the speed of the planes very materially.

There were also fundamental differences between the British and American development of radar. Radar in England was developed mainly under the auspices of the Royal Air Force, and in America under the auspices of the Navy through the Naval Research Laboratory. The radar sets used aboard ship in the Royal Navy were nothing but adaptations of airborne and shore equipment, whereas our surface-ship radar equipment was designed specifically for that purpose. When we disclosed our submarine radar to the British, they had none. Neither the British nor the Americans had fire control radar at this time.

One of the inventions which the British disclosed to us was a great contribution to radar—an oscillator tube of unusual power output, known as the multi-cavity magnetron, invented at the University of Birmingham in England. As a matter of fact, in order that the work of the Naval Research Laboratory may be properly and historically evaluated, let me state that no magnetron sets were installed in the Royal Air Force at the time of this disclosure, and they were not installed until late in 1942 by both the British and the Americans almost simultaneously.

There has been so much misinformation published about this magnetron that it is just as well to get the story straight for once. The Naval Research Laboratory was using magnetrons as early as 1933 in an attempted development of CW (continuous wave) radar, but they were not cavity magnetrons and were not suitable for pulsing. The vastly improved magnetron developed in England was of course the cavity type with a heavy cathode suitable for high power pulsing. In "Radio Age" (Radio Corporation of America), October 1947, Dr. J. S. Donal, Jr., states: "Not only did this tube make possible the radar warning system credited with saving England in the critical days of the blitz but the numerous later forms of radar, which proved invaluable to all branches of the armed services, owed their effectiveness in large part to the magnetron." That statement of Dr. Donal is just not true.

Let us look at the record. The Battle of Britain, sometimes referred to as the "Blitz," took place in 1940. The first attack occurred on July 10 and the final phases occurred in December of that same year. The radar warning system credited with saving England in those critical days operated on a frequency of 40,[5] not 3,000, megacycles, using grid-

[5] The original British air warning system was known as CH (Chain Home). It operated within the range 20 to 60 megacycles per second.

controlled triodes, basically not different from the commercial transmitting triodes of that day. The airborne radar which figured in the later stages of the Battle of Britain also used grid-controlled triodes. These tubes were disclosed to us by the British in the summer of 1940, and an attempt was made to interest our manufacturers in their production. The tubes operated at a frequency of a few hundred megacycles. As late as May 1941, in the operation which led to the sinking of the *Bismarck*, which was equipped with fire control radar, only two British ships—both cruisers—had radar, one set of limited use and the other good for 13 miles.[6]

This estimate is confirmed by a statement of Dr. E. G. Bowen of Sir Henry Tizard's 1940 mission when he said:

". . . pulsed radar detection equipment appeared in the United States at about the same time as in England. A remarkable feature . . . was that in spite of the fact that there was no communication between the participants, there was striking similarity in the final equipment. The first American air warning equipment for use on board ship operated on 200 megacycles per second, a frequency identical with that of a British ground-based equipment. The aerial system consisted of the same number of elements mounted on a similar reflecting mattress, while the transmitter and receiver were almost indistinguishable in each case."[7]

In this country, land-based search radar operated in the range of 100 megacycles, shipborne search radar and land-based fire control radar at 200 megacycles, and airborne search radar at about 500 megacycles, all using grid-con-

---

After the heavy German daylight attacks on Britain ceased in 1940 and the enemy began night bombing raids, the British "countered by the development of GCI (Ground Control of Interception) and AI (Aircraft Interception) techniques. . . . The equipment developed for GCI operated on 200 megacycles per second and with modification it served during most of the remainder of the war."

A Textbook of Radar, by E. G. Bowen, Angus & Robertson, Sidney, London, 1947, pp. 513, 520.

[6] *The Bismarck Episode*, by Captain Russell Grenfell, RN, The Macmillan Company, London, 1949, p. 45.

[7] *A Textbook of Radar*, by E. G. Bowen, p. 3.

trolled triodes developed by the Eitel McCullough Company at San Bruno, California. Our shipborne fire control radar, FC, at that time (1940-42) operated in the region of 700 megacycles, using grid-controlled triodes developed by the Bell Telephone Laboratories.

The cavity magnetrons brought to us by Sir Henry Tizard's mission were early models and further developments took place subsequently. The British did have a set, however —a simple "A" scope type search set—suitable for surface search purposes and using this early version of the cavity magnetron. One such set was seen on a British corvette in New York harbor in the summer of 1941. It was more of a laboratory version than an engineered service equipment.

The announcement of this new high power magnetron tube did provide, however, a great stimulus to the further refinement and development of the radar principle. With the very high frequencies made possible by the magnetron, sharply directive and powerful beams of radio waves could be had with the use of antennas which were small enough to be easily mounted in any desired location on a Naval vessel. These antennas were even installed in airplanes in such a position that they could be rotated through a full 360° without too much difficulty. It can be seen that this introduced a flexibility in the use of radar which instantaneously called for its development along several different but parallel lines to answer the various problems brought up by its projected uses.

The promise held out by the engineers responsible for the development of the early centimeter radar more than came true in its first use in the Fleet. Consequently, an enormous amount of effort was spent in developing many different radar systems, each of which was tailored for a specific operational need.

The introduction of centimeter radar had a great impact on the second half of the war in Europe. While it found many uses here, its principal advantage came about in its use in AI (Air Intercept); that is, intercept sets which permitted a fighter plane to search out and destroy an attacker in darkness. The second important use of centimeter radar in the European Theater was, of course, its use in anti-submarine warfare.

The Germans perfected a technique of intercepting our planes equipped with radar by using a search receiver. As we went up in frequency, our planes became more effective because the Germans failed to anticipate our ability to change frequency in radical steps and were slow in converting their search receivers.

By the time we traded information with the British, the Naval Research Laboratory had introduced into the search radar system, as previously noted, most of the important parameters and variables that exist in radar as we know it today.

The idea that the British, by the time of the Battle of Britain, were completely equipped with radar is wrong.

One of the things that surprised the British when we were making our mutual disclosures was the fact that we took the contractors into such complete confidence. Too often in England, at least at that time, it was the custom with a radar set or other secret material to obtain the component parts from several different manufacturers or contractors and to complete the assembly in a government plant. The British were amazed to find that in all of our secret equipment the entire article was produced by one company. I explained to them that this procedure had been very successful in time of peace and was particularly adapted for emergencies and for the great requirements of war production. I further explained that in time of peace between the wars anyone could have gone to the Bell Telephone Laboratories, for instance, and seen a whole wing or corridor set aside for secret government work with a company policeman on duty to deny access to the area to anyone who did not have the proper credentials.

As the British excelled us in the beginning in airborne radar—a superiority which we soon overcame and maintained—so we excelled them in the beginning in regard to surface-ship and submarine radar. It was impossible to convince Secretary Knox, in the early stages of our collaboration with the British or at any time, that we not only had many more ships equipped with radar than the British but that ours were immeasurably superior because they had been designed specifically for ships' use. My estimate of the quality and quantity of ships' radar employed by the Royal Navy is confirmed in Captain Grenfell's story of *The Bismarck Episode*. The search for and destruction of the *Bismarck* took place 22 to 27 May 1941, almost a year after our disclosures to Sir Henry Tizard.

In October 1940, in addition to my duties as Director, I was designated as coordinator of all phases of the Navy's radar program.

I was so much encouraged by the interest developed on the part of the General Electric Company that on 16 October 1940 I appealed to Mr. A. W. Robertson, Chairman of the Board, and Mr. George Bucher, President, of the Westinghouse Electric Corporation, to participate in the Navy's radar program. One day we were visited by a party of Westinghouse representatives headed by Mr. George Bucher and including Mr. Walter C. Evans, Manager of their Baltimore Radio Plant, and Dr. John A. Hutcheson and Mr. C. J. Burnside of the same activity.

It so happened that the work of converting the Laboratory altimeter into an airborne search radar operating at 500 megacycles was well underway, and when Mr. Bucher became so impressed by possibilities of radar that he expressed his willingness to engage immediately in radar activity I suggested that the only way he could catch up with the Naval Research Laboratory was to start building radar, specifically this airborne set (ASB).

As a result of this conference, Westinghouse, which had never built any radar, accepted a contract to build this airborne set. By October our pilot model was far enough along for the Westinghouse Electric Corporation to start production models. By November the Radio Corporation of America was brought into the situation and, in 1942, Bendix. This equipment was known as ASB and was first produced by Westinghouse in May 1942. It was tremendously successful in extending the area which could be covered by a patrol plane and therefore was of great assistance to the anti-submarine program. It was also very useful for navigational purposes. Over 26,000 sets were produced with a considerable number of them turned over to the Army and to the British, since it was actually a tremendous improvement over the British airborne radar. Ultimately these sets were replaced by sets of still higher frequency, but many were still in operating use by the time the war ended. They were especially valuable at this time because sets of much higher frequency were absolutely unobtainable.

As early as July 1939 Western Electric had demonstrated to the Army and Navy an experimental equipment designed to meet fire control problems. It operated in the range of 500-700 megacycles which was a much higher range than had been used before. It employed an antenna about six feet square. Both the Army and Navy ordered a prototype model equipment CXAS. In December 1940, after a few tests, the Bureau of Ships ordered ten and decided to standardize this class of application on the 700 megacycle band. The first of these FC's, Mark I, was installed on the *Wichita* early in July 1941. Quantity production was in progress by the time of Pearl Harbor. Eventually almost 1,000 of these equipments, of various Marks, were manufactured and installed in our ships. Almost 400 were still in the fleet on V-J Day. One advanced design followed another in rapid succession. As they improved, they featured many of the essentials in radar systems which were incorporated as soon as they appeared.

Mark II was similar to Mark I but embodied continuous tracking in range and azimuth and integrated the previously existing optical system with the new radar system. Before this Mark II got into production a much higher powered transmitter was developed and with this change the equipment was renamed Mark III.

". . . at the beginning of 1941 there were still many unsolved problems in 3000 megacycle radar other than that of the transmitter tube. On the other hand, the system's problems had been quite satisfactorily solved in the 700 megacycle region. The decision was therefore immediately made to extrapolate the British design (multi-cavity magnetron) down to 700 megacycles in order to obtain a higher powered and more satisfactory oscillator for the existing systems. . . . This was the first type of multi-cavity magnetron to go into production in this country."[8] Concurrently, "Bell Laboratories developed an improved tetrode modulator which was many times as efficient for radar pulse service as the triodes formerly used."[9]

Soon after the deliveries of Mark III (with magnetron) began, Western Electric successfully incorporated the General Electric lighthouse tube, mentioned later, to reduce receiver noise. Mark III was installed on the battleships *Pennsylvania* and *New Jersey* and the cruisers *Honolulu* and *San Francisco* at an early date.

Mark IV (also with magnetron) was developed for use with 5" anti-aircraft guns. This radar was demonstrated on the destroyer *Roe* in September 1941 and deliveries began in December 1941.

Mark IV was installed early on the battleship *Washington* and on the aircraft carrier *Saratoga.*

The development of the so-called "lighthouse" tube for radar illustrates the difficulty which laboratories encounter in time of war. The General Electric lighthouse tube, a typical triode, was a disc seal type tube using plane parallel electrodes whose leads were circular discs projecting straight through the glass envelope.

General Electric engineers disclosed drawings of this tube to the radar personnel of the Naval Research Laboratory. They had never been able to obtain company approval to proceed with its development. The Naval Research Laboratory then entered the situation and gave General Electric a development contract to produce the tube. The transition from manufacture by hand to manufacture by machine proved

[8] *Radar Systems and Components,* Bell Laboratories Staff, D. Van-Nostrand Company, Inc., New York, 1949, p. 44.

[9] *Loc. cit.*

to be exceedingly difficult. The tubes were urgently needed afloat and it was almost impossible for the Naval Research Laboratory, where the development of the tube had been initiated from the drawings supplied by General Electric, to obtain anything except rejects for further experimentation.

As early as 1940 and 1941 the Naval Research Laboratory had put into production and installed on most carriers high-frequency homing equipment. The importance of homing equipment for carriers is obvious. These were in use as early as the Battle of Midway, 4-7 June 1942, and in the African invasion which began on 7 November 1942. Pilots navigate to their targets but frequently lose track of their positions in the course of dogfights or evasive flying to avoid anti-aircraft fire. It is a great comfort for the returning pilot, with fuel running low, to home in a bee-line on the carrier's radar beacon (or racon as it was called). Many pilots and planes were saved by this device.

A cruise of the *California* from Bremerton, Washington, to the vicinity of San Pedro, California, proved that navigation in dense fog at high speed was safe and practicable with radar.

In 1940 it was observed that, under certain atmospheric conditions, which were not understood at that time but which were fully explained later, radar reflections could be obtained from mountains as far away as 450 miles.

In 1940 the Chief of Naval Operations had estimated that the radio ranging equipment procurement program which was being initiated would require ten million dollars in 1941 and twenty million dollars in 1942.

The Naval Research Laboratory did considerable work at an early stage on radio recognition equipment, IFF (Identification, Friend or Foe), and ultimately this work was enormously expanded and the Laboratory became the national center for this type of enterprise under Dr. Claud E. Cleeton.

Because of the one-sided presentation of radar to the public in this country, the public by and large unquestionably had the feeling that we had obtained all of our radar from Britain; and this view still obtains with some uninformed commentators, magazine writers, and authors. This situation got so bad that I was advised, unofficially, by the Commissioner of Patents, Conway P. Coe, to see what I could do to counter this false idea through the daily press. Fortunately, Lieutenant Commander Newell Atwood, USNR, whom I had placed at the Naval Research Laboratory as its first patent lawyer in March 1942, had already worked out a very satisfactory chronology of the history of radar discoveries at the Naval Research Laboratory. Through the collaboration of the Laboratory and myself with Mr. John Hightower of the Associated Press, Mr. Hightower was able to present the American side of the radar picture in seven articles through the medium of the Associated Press. These articles had the second largest circulation ever obtained by an Associated Press article. They ranked only behind an article on "The Forgotten Man" published after World War I.[10] Even now I am amazed that I was able to get the necessary clearance from the Navy Department. After the appearance of the Hightower articles, 21 through 26 May 1943, the curtain of secrecy was again drawn over the radar situation.

One of the early developments of the Naval Research Laboratory was the sono-radio buoy. This buoy carries a microphone for picking up noises such as propeller sounds in the water, the microphone being suspended from the buoy. An antenna is affixed to the top of the buoy for transmitting by radio those noises which are picked up by the microphone. Transmission is effected for moderate distances to ship, shore, or airplane. These buoys were intended for harbor protection and for advance bases. The buoy was small enough to be carried by plane and could be dropped in any area which it was desired to search by this method. The National Defense Research Committee did an excellent job in getting this buoy into production, and the buoys, as produced, were exact copies of the Naval Research Laboratory pilot models.

[10] By arrangement with Senator James G. Scrugham I procured the publication of the Hightower articles as a public document, No. 89, U.S. Senate, 78th Congress, 1st Session, 2 July 1943.

They were widely used during the war and were also used in Operation Crossroads at Bikini for closely observing underwater effects and transmitting the results by radio to recording equipment and observers who were located at a safe distance from the scene of the explosion.

Chesapeake Bay Annex. While all of this work was going on in radar, it became obvious that the dream of Dr. Taylor and Mr. Young to construct a branch laboratory for the test of radio and radar equipment over water, reasonably distant from Washington and near Navy air fields, should be realized. At the same time there was underway a terrific expansion in the Naval Research Laboratory—that is, a terrific expansion of its area and its facilities. When I went to the Naval Research Laboratory in the fall of 1939, the highest building number was 11, and when I left in the fall of 1942 the highest building number was 41. By that time I had also added the Chesapeake Bay Annex, which realized the dream of Taylor and Young. Incidentally I might add that, during this period, the radar section of the Laboratory increased from six to three hundred individuals.

The National Defense Research Committee offered to finance the Chesapeake Bay Annex, but I felt it very much better to do it with our own money. The Annex was first conceived on a very modest scale. However, I took the precaution to obtain plenty of land, although even this amount which seemed so much at the time was later added to, principally for security reasons. Taylor and Young envisaged a spot where transmitting and receiving apparatus could be set up for over-water operations. The housing facilities were to be only sufficient to handle week-end or short visits from Laboratory scientists and personnel. No more equipment would be taken down than could be handled by the ordinary Research Laboratory radio and sound truck.

The work had no sooner started than more ambitious plans were underway and before preceding plans had been completed. The Bureau of Ordnance early participated in the development of the Annex, and in no time at all the Annex bore the same relation to its original conception as the oak bears to the acorn.

I remember, on returning from one visit to the Chesapeake Bay Annex where the Appropriations Committee of the House of Representatives had the opportunity to inspect the work we were doing in radar, that Representative Plumley of Vermont offered me an immediate increase of one million dollars in Laboratory funds. I told him that research work had this peculiarity: funds spent for research are mostly for labor, very little for material as a rule, and our scientific force could not be built up rapidly enough to spend that additional one million dollars in one year.

The Army, the National Defense Research Committee, the Bureau of Standards, and many other Government agencies found this Chesapeake Bay site extremely valuable for operational and engineering evaluation tests. Frequently, contractors would have twenty or thirty of their engineers and scientists housed and fed at this site while equipment which they had developed under contract was undergoing tests and improvement.

Under the plea that the National Defense Research Committee might grab off the Naval Research Laboratory, "it was decided" to transfer the Laboratory from the jurisdiction of the Secretary of the Navy to the Bureau of Ships. This was done by General Order #150 dated 12 July 1941. That may have been the truth or it may not have been the truth, but it operated to deprive me of "free gangway" to the Secretary's office, and that may have been the real reason. The Secretary of the Navy at the time was Knox. It did make this real difference—all plant additions had to be submitted via the Bureau of Ships.

As long as the Research Laboratory's projects for expansion were submitted through the Bureau of Ships, the Laboratory and these projects did not rank as high as Bureau level and could easily be presented by the Bureau to the Shore Development Board at such a low priority as to virtually insure their rejection before they were submitted to the Budget. At the particular time of which I am speaking, Admiral Ben

Moreell was Head of the Shore Development Board. By arrangement with him all of our projects were placed as high as any projects submitted by any Bureau so that, in effect, this arrangement by-passed the Bureau of Ships. This permitted me to get my projects approved by the Navy Department Budget Office and up to the Budget without any interference at all. The presentation of these projects to the Congress was particularly easy because the Naval Appropriations Committee of the House of Representatives was always interested in furthering the progress of the Naval Research Laboratory and its many projects.

Pearl Harbor. The disaster at Pearl Harbor was a terrible shock to those of us at the Naval Research Laboratory as well as those of us who had spent our entire lives trying to prepare the Navy for the war with Japan and Germany which we had long been convinced was inevitable. We materiel people felt that the military had let us down. Some of our best efforts were at the bottom of Pearl Harbor. And what about radar?

There had been constant collaboration between the Naval Research Laboratory and the Signal Corps Laboratory at Ft. Monmouth ever since the Naval Research Laboratory had made its first disclosure to the Army. The fact that the first Japanese planes over Pearl Harbor were located by the Army's radar and the information not properly used seemed to us to be tragic. Dr. Taylor was particularly dismayed, feeling as if all his lifework had been thrown away. Of course we had radar on certain ships at Pearl Harbor at that time, but they were ineffectual because of the surrounding hills and mountains which blanked off their searching capabilities.

We had had the greatest difficulties imaginable in introducing radar, in getting its potentialities realized, and here was the most wonderful opportunity in the world for radar to demonstrate itself in the most dramatic fashion imaginable. This opportunity had been lost because the Army at Pearl Harbor didn't have the faintest idea of the possibilities of radar. I never dreamed that anybody at that stage of world affairs would not have kept a continuous night and day radar watch, all the way up to the top command.

At the time of Pearl Harbor I was performing dual duty as Director of the Naval Research Laboratory and as Officer-in-Charge of the Federal Shipbuilding and Dry Dock Company at Kearny, New Jersey, which I had seized and was operating under an Executive Order of the President of the United States. By choice I was wearing uniform because, during this—my first seizure and operation of an industrial plant—I was very anxious to use every device I could think of to set myself apart from both labor and management. The adverse comments on the military that I heard on the trains between Washington and Newark burned me to the quick because I agreed absolutely with the ordinary citizen on the disaster.

After Secretary Knox removed Admiral James O. Richardson from the command of the U.S. Fleet at Hawaii before Pearl Harbor, he had offered the command to Rear Admiral Chester W. Nimitz who was then Chief of Personnel. Nimitz, with his usual self-abnegation, had told the Secretary that there were so many people senior to him on the lineal list who were competent to command the Fleet that he preferred to wait until his turn came.

I was very much surprised when Admiral Husband E. Kimmell was selected to relieve Richardson. Kimmell was able, hard-boiled, and brusque, a type which usually has a hard time in getting to the top. I was shocked, after the news had been received of the debacle, to learn that Kimmell had kept so many ships—more than one-half of his Fleet—in Pearl Harbor at one time. Of course, I knew that Kimmell had demanded of Secretary Knox more air cover for his Fleet and that he had been rebuffed by the Secretary who had said that the planes had to go to England and he couldn't have any more.

The Congressional hearings on Pearl Harbor meant nothing to me at all. Kimmell was as fine and able a man as I ever met. He and General Short were the goats. No one seems to know that the Army's mission in Hawaii was to defend Pearl Harbor.

Mr. Knox felt as if he had been terribly let down by the Navy after he received the news of Pearl Harbor. I will say for Mr. Knox, whom I shall criticize later for other things, that no man in the national government realized better than he that we were approaching war. He cited that fact at every meeting of the Council of the Secretary of the Navy. I still cannot understand how we got caught in the surprise raid of the Japanese at Pearl Harbor. All my generation of Naval officers were taught never to forget that Japan started the Russo-Japanese war with a sneak attack on Port Arthur, and we accepted the idea that war with Japan would start with a sneak attack on us. Our repeated rebuffs to Japan before Pearl Harbor plainly indicated that the only outcome of the diplomatic impasse in which we had allowed ourselves to be placed was war.

Responsibility begins at the top. Everyone in the high command in Washington, from the Commander-in-Chief, i.e., the President of the United States, down was familiar with the dispositions at Pearl Harbor and, presumably, approved of them. They had the power and the responsibility to order them differently if they did not approve. For years we had joked in the Navy about what would happen if the enemy had ever attended our own Naval War College. Well, the enemy did what we didn't expect him to do. The attack on Pearl Harbor was not logical; he had not attended our War College and, therefore, the totally unexpected happened.

While I was Engineering Chief of the Navy from 1935 to 1939 I never expected to complete my term before the United States was engaged in the irrepressible conflict. I made extensive preparations at the Naval Research Laboratory for war, stock-piled every strategic material I could think of and stored a year's supply of coal for our power plant just in case. When I seized Federal Ship, I organized air raid squads and overhauled and put into commission fuel oil storage which had not been used for years.

At the time of Pearl Harbor the major units of the Fleet

were equipped with what has now come to be known as early warning radar after a design by the Naval Research Laboratory, and installations were being made of an early fire control radar which was developed by the Western Electric Company after the basic ideas behind the development of radar were revealed to them by the Naval Research Laboratory personnel. In the years just preceding Pearl Harbor, some parts of the Pacific Fleet had been using radar as part of the doctrine in their fleet practices with aircraft. In addition to the use of radar in the fleet practice, it was also used in maneuvers, especially for operations under darkened ship conditions and as an aid to navigation in returning to the base at Pearl Harbor.

As of 10 July 1942, there were 524 radar sets installed aboard naval surface craft and 1,544 sets installed aboard naval aircraft. There were approximately 300 radar sets aboard surface craft and a similar number aboard aircraft for the primary purpose of detection of enemy surface and aircraft. All but about 30 of the shipborne detection radar sets were duplicates of equipment originating at the Naval Research Laboratory and furnished to commercial concerns as "prototypes." The airborne equipment was largely patterned after the British design but was rapidly displaced by the Naval Research Laboratory model ASB. Both the shipborne and airborne detecting equipment operated on approximately 200 megacycles. There were some detection sets operating at 3,000 megacycles installed on surface craft.

As of the same date, there were approximately 200 fire control radar sets installed aboard naval vessels operating on 700 megacycles.

Also as of the same date, there already had been installed on naval vessels 50 identification radar sets, all built from prototypes originating at the Naval Research Laboratory, while aboard naval aircraft there were approximately 1,250 identification sets. This equipment was designed to operate with the detection radar equipment and, in general, also operated at approximately 200 megacycles.

From the above it is apparent that the Navy's fighting

forces at that time were working for the most part with radar equipment operating in the frequency range so comprehensively studied at the Naval Research Laboratory, namely 100-500 megacycles. It is furthermore apparent that, at this time, the widespread installations of radar equipment in the fleet were patterned after Naval Research Laboratory prototypes.

Still as of 10 July 1942, orders had been placed for 10,000 additional shipboard radar sets and 130,000 aircraft sets.

Training of Personnel in Radar. The training of personnel to operate and maintain radar equipment installed aboard naval craft was a real problem. The Naval Research Laboratory contributed immeasurably to the solution of this problem through its Radio Material School located on the Laboratory grounds. Here enlisted men and officers of both our Navy and the British Navy were given intensive training in the operation and maintenance of radar equipment identical with that installed in the fleet. This training could not be surpassed elsewhere because there were available at the Laboratory not only the scientists and engineers who had invented and developed radar but because the latest developments in the field were made available immediately to the students in the school. The value of this training was clearly indicated by the fact that, in addition to the regular Navy personnel training in the school, a large number of civil service engineers were given an intensive course in radar at the school at the request of the Navy Department. These radar graduate engineers proceeded to perform procurement, consulting, and maintenance work in the Navy Department and at nearly every United States Naval Base within and without the continental limits. A total of 1,195 men were trained in radar at the Laboratory school by 10 July 1942, as shown by the following tabulation:

U.S. Navy officers	371
U.S. Navy enlisted men	748
British Navy officers	4

British Navy enlisted men	8
Civilian engineers	64

In addition, the staff of the school established two more radar schools at San Francisco, California, and Corpus Christi, Texas, for the training of over 600 radar men monthly.

The dependency of the commercial contractors upon the scientific and engineering knowledge of the Laboratory personnel reflects the contribution of the Laboratory to radar development. This dependency is shown by the following list, indicating the visits made by engineers of some of the companies engaged in radar production to the Naval Research Laboratory for consultation with Laboratory engineers during the month of March 1942:

Hazeltine Corporation	15
General Electric Company	26
Radio Corporation of America	36
Westinghouse Electric Corporation	36
Philco Corporation	34
Collins Radio Company	10
Bell Telephone Laboratories (April)	8

Not long after Pearl Harbor our naval forces became engaged with the enemy in the South Pacific where radar became evaluated and integrated into naval tactics.

Battle of Coral Sea, 7-8 May 1942. Japanese carriers at this time did not have radar or homing devices, both of which our carriers possessed. Furthermore, "the American radio telephone inadvertently jammed the frequency used by their (Japanese) aircraft preventing the pilots from getting a bearing on their carriers"[11] so that the Japs had to use searchlights for night homing. The *Lexington's* CXAM, built by the Radio Corporation of America to Naval Research Laboratory models as previously described, detected the first sign of the attack on May 8th—Jap planes 70 miles away. In this

[11] *History of U.S. Naval Operations in World War II*, Samuel E. Morison, Little Brown and Company, Boston, 1950, vol. 4, p. 44.

battle, the *Yorktown* was also equipped with CXAM, but the destroyers as yet had no radar.

Battle of Midway, 4-7 June 1942. There were two good search radars (SCR-270) installed on the island of Midway, but most of the aircraft based there did not have radar. The *Enterprise* had a CXAM and the *Hornet, Yorktown,* and a few of the cruisers had SC sets which, it will be remembered, were manufactured by General Electric from Naval Research Laboratory models. "Our carriers were equipped with automatic homing equipment, and several of . . . [our] . . . planes used it successfully to locate *Enterprise.*"[12] Records indicate radar detection of enemy planes, some of which were as follows: Midway Island (SCR-270) picked up planes at 93 miles distant; *Yorktown* at 40 miles, and also warning of plane attack; and a cruiser (SC) the second attack wave at 40 miles.

Battle of Savo Island, 9 August 1942. The destroyers *Blue* and *Ralph Talbot* were equipped with SC radar and they were assigned to patrol duty. Morison says "the two destroyers could be as much as 20 miles apart at the outer end of their patrol lines, a situation which actually occurred when the enemy struck; and their SC radars were reliable only up to 10 miles under optimum conditions, which (owing to nearby land) did not here obtain."[13] This comment is not understood unless the reference is to surface vessels. We have just noted a few lines above the detection by a cruiser with an SC set of an air attack wave 40 miles distant. The cruiser *Astoria* was equipped with fire control (FC) radar manufactured by Western Electric under a Bureau of Ships contract. The *San Juan* was equipped with the new magnetron microwave search radar (SG), a great improvement over the earlier SC, perhaps its first appearance in this area. The *Ralph Talbot* also had an automatic gun control system.

Battle of the Eastern Solomons, 24 August 1942. In this battle we note that radar was detecting planes at 100 miles (*Enterprise*), 188 miles, and so forth. It was also observed

[12] *Ibid.*, Vol. 4, p. 130. [13] *Ibid.*, Vol. 5, p. 30.

that the electronic identification system (Naval Research Laboratory) was at that time "far from perfect." There were lots of things that weren't perfect at that time.

Submarine Patrols. For the sake of chronology it is just as well to interrupt the recital of the use of surface radar in the series of these early naval battles and take a look at what the submarines were doing in radar and in sound detection.

At first, before the Naval Research Laboratory radar was installed, the submarines used the periscope and the underwater ultra-sound system for target detection. Morison says that ultra-sonic systems "never lived up to pre-war expectations."[14] There will be more said about sonar later. During 1942, submarines joined the fray with the Naval Research Laboratory search radar. In August and September 1942, the *Haddock*, a new submarine, appeared with an SJ search radar, designed principally for the fire control function, built around the magnetron, and manufactured by Western Electric. These two search radars proved to be very useful.

Battle of Cape Esperance, 11-12 October 1942. The Japs still had no radar. The *Helena* and *Boise* were equipped with SG radar. The *San Francisco* had fire control (FC) radar. This battle showed that radar was not yet worked into battle tactics.

On 16 October 1942 the *South Dakota* with FC radar was credited with shooting down 38 out of 38 planes. This marks the first use of the proximity fuse, at least in this area.

Battle of Santa Cruz Islands, 26-27 October 1942. The presence of the Japanese battleships *Kongo* and *Haruna* in this area marks the first appearance of Japanese radar. Both ships were equipped with search radar.

The *South Dakota* was active here with her radar, as was the *Boise*, equipped with FC radar.

On 4 November 1942 the *South Dakota* with her Mark III (magnetron) FC was credited with sinking a major Japanese vessel eight miles away with two salvos.

[14] *Ibid.*, Vol. 5, p. 93.

Battle of Guadalcanal (Naval), 12-15 November 1942. Our forces received ample warning of the opening attack from radar and coast watchers. The Flagship *San Francisco* had fire control (FC) radar for surface and anti-aircraft. Three cruisers and two destroyers were equipped with the latest SG radar and should have been in the van instead of the *Atlanta* with older (SC) radar. The *Helena* with her SG picked up two groups of enemy vessels at 27,000 and 32,000 yards respectively. Japanese Naval forces under Admiral Abe did not have any radar. The destroyer *Fletcher* had an SG. The *Washington's* radar picked up a target nine miles away.

Morison, in describing this battle, says: "Faith in radar had become so great that the loss of both SG and SC equipment had a most depressing effect on officers and men; it gave all hands the feeling of being blindfolded."[15]

Battle of Tassafaronga, 30 November 1942. Each of one destroyer group and two cruiser groups included at least one ship equipped with SG radar. The Japanese were again at a disadvantage in that their forces in this battle had neither radar nor means of monitoring American radio circuits. Morison comments: "Radar operators became the most important individuals in the force as, with hands on dials and eyes glued to the scope, they endeavored to sort out ghost-like blobs indicating 'pips' that might be ships."[16] Actually, radar picked up the enemy almost dead ahead at 23,000 yards. Again Morison: "As at the Battle of Cape Esperance, the Americans had beaten the Japanese to the draw by using radar. . . ."[17] The *Pensacola* had fire control (FC) radar and the old SC radar.

Operations Around Guadalcanal, 1 December 1942 to January 1943. Radar became indispensable to the Catalina flying boats in their night operations.

Battle of Rennell Island, 29-30 January 1943. IFF was not too dependable but fire control radar was excellent. The flagship picked up enemy planes at 60 miles.

[15] *Ibid.*, Vol. 5, p. 276. [16] *Ibid.*, Vol. 5, p. 297. [17] *Ibid.*, Vol. 5, p. 300.

Thus we see that the early pivotal battles of Coral Sea and Midway were fought exclusively with Naval Research Laboratory radar. The first SG set made its appearance at the Battle of Savo Island. Apparently only Naval Research Laboratory radar was used in the Battle of the Eastern Solomons. Two SG sets are mentioned in the Battle of Cape Esperance. In the Naval Battle of Guadalcanal three cruisers and two destroyers were equipped with SG, but this type of radar was still scarce in the Battle of Tassafaronga. During all these battles in 1942, the only fire control radar used was Western Electric's 700 megacycle, in successively improved models which stemmed from the original contract of 1940.

From the evidence gathered from captured Japanese equipment it was obvious that our fleet was better equipped in 1942 than the Japanese were even at the end of the war. One of these early Japanese sets was captured at Guadalcanal and was shipped to the Naval Research Laboratory where it was assembled and operated. Later a much improved set, very likely made in the latter part of 1942, was obtained with the capture of the islands of Attu and Kiska. "When the Japanese took the Philippines, they gained access to some very good search radars. They captured another one of an excellent British design at Singapore, but it had been well broken up and they could not make much of it. Unfortunately, however, they discovered somewhere a copy of the instruction book and so were able to build it although they never did succeed in making a copy of it as good as the original."[18]

The whole period of early radar development was delayed somewhat by lack of vision. Before an airplane interfered with reception, in the case of Hyland's experiment in 1930, Taylor was all for turning his work over to the Army and forgetting it because he saw no naval application for reasons that have been explained. Western Electric, General Electric, Radio Corporation of America, and Westinghouse were

[18] *Radio Reminiscences: A Half Century,* Chapter xv, "Science at War," by Dr. A. Hoyt Taylor, p. 378.

familiar to some extent for years with the work at the Naval Research Laboratory, but they took no active part in this work because they saw no commercial application. The records of the Naval Research Laboratory reveal that when its developments were disclosed, as I have said before, to a group of engineers from the Bell Telephone Laboratories, the engineers said: "Thank God you are doing this work, because our Board of Directors would never permit us to do it." The Radio Corporation of America furnished the first contract sets for the Navy in 1940 and thus had the opportunity to dominate radar for years to come.

While the Naval Research Laboratory, the father of radar in this country, was established in 1923, pursuant to the recommendation of the Naval Consulting Board whose Chairman was Thomas A. Edison, the Laboratory grew very slowly, due more than anything else to the Navy's failure to recognize the possibilities of research. After all, the rapid rise of the industrial research laboratory in the United States did not take place until after, and partly as a result of, World War I, so perhaps the Navy did not do so badly at that. Before World War II no industrial research laboratory had the capacity or the talent to have taken over the bulk of radar development. That required a state of war and an electronic effort comparable to that put forth by the Office of Scientific Research and Development and the many electronic companies.

The Navy of course was at a disadvantage in that it was not organized to handle research on a broad scale until the organization of the Office of Research and Inventions in the spring of 1945, shortly before V-J Day. This office was given a statutory existence by Congress in the summer of 1945 as the Office of Naval Research. It is interesting to note that my proposal to set up a central research agency in the Navy Department, made in the spring of 1941, was reported on adversely by my old antagonists, the General Board of the Navy, after hearings, and the adverse report was concurred in by the Secretary of the Navy. The tip-off was that the General Board and Secretary Knox saw no virtue in my proposition and thought it represented just another attempt of the same Admiral Bowen to build an empire.

The development of radar at the Naval Research Laboratory was delayed somewhat by the inability to obtain the right type of personnel. A television engineer had an ideal background of education and experience for radar work, but such engineers were few and far between at that time.

What was the scientific world in the universities doing during this period? They performed nobly under the National Defense Research Committee and the Office of Scientific Research and Development at a later date, but I am not aware that they made any contributions to radar during the pioneer years. We were in touch with the National Academy of Sciences, but that contact did not extend to those scientists who later constituted the Radiation Laboratory of the National Defense Research Committee.

It was noticeable that the National Defense Research Committee representatives failed to maintain a close contact with the Naval Research Laboratory and to learn from the years of experience of our personnel. When the National Defense Research Committee began to assemble the large group of scientists which formed the Radiation Laboratory they coralled a large number of individuals from industry and from the universities. These individuals were undoubtedly fully abreast of all the developments in the art of electronics *which had been published,* but the Navy's policy of keeping secret its work on radar meant that the National Defense Research Committee personnel were neither aware of nor in a position to place any value upon the work that had been done by the Navy prior to 1940. There was, no doubt, a natural feeling on the part of this personnel that they knew all there was to know about radar at the time they were drafted, so to speak, to work on radar. While the Navy's disclosures no doubt gave them a broad outline of possibilities, they probably accepted this information as a statement of their problem, the solution of which was left to them. That might account for their failure to make frequent visits to the Naval Research Laboratory or to seek, at the Naval Research Laboratory, detailed information. This possibility is strengthened by the fact that frequently disclosures in Radiation Laboratory reports of developments new to the authors of these disclosures were anticipated by identical and earlier disclosures in Naval Research Laboratory reports and documents antedating the formation of the Radiation Laboratory.

Any description of engineering and science in the Navy must stress the point that a great deal of valuable scientific work can be and is done within the military services which cannot, for security reasons, be published. In other words, the total store of scientific knowledge has a secret compartment, the existence of which must be recognized (even though the contents are unknown) and effectively utilized in any mobilization of science in time of emergency. In this country the usual training of a scientist or engineer leads him to believe that everything one needs to know regarding a certain subject is available in the literature of the subject and that no further look is necessary when undertaking an assigned problem.

The National Defense Research Committee, being well loaded with university professors, was able to mobilize all the members of the union, something which the Naval Research Laboratory never could have done in a thousand years. Also, the National Defense Research Committee had a private pipeline to the President of the United States and the U.S. Treasury.

Nevertheless we must be practical. I do not think, for reasons I have cited, that science could have been mobilized for use in World War II by any other method than the one used, and those of us who had been working in applied science for years cannot be blamed for not always enthusiastically endorsing all the efforts of the Johnny-come-latelys who inevitably steam into Washington at the beginning of a war.

It is pertinent to call attention to the fact that these Office of Scientific Research and Development scientists did not work as scientists but as engineers, and only the war emergency could have induced them to have played such a role—an invaluable one in the National effort.

It took a long time to sell the Navy on radar. Remember, the Bureau of Ships, not the Bureau of Ordnance, let that most important first contract for fire control radar. One admiral didn't want the first radar sets put in his ships because they would spoil the ship's appearance. The first antennas, as previously indicated, were known as "flying mattresses" because of their resemblance to gargantuan mattresses. Admittedly they were not beautiful. On one cruiser we learned that the Commanding Officer and the Executive Officer never used this new radar. They said they hadn't asked for it and didn't want it. Another distinguished admiral, one of our most famous early flyers, shrugged off the possibilities of radar by saying to me that planes could avoid radar by flying low. Yes? Some of this lack of interest plainly reflects a lack of information and a lack of indoctrination. No one in his senses will neglect anything that may some day save his life. But after all the cold reception of radar is characteristic of the reaction to almost all great inventions.

It is true that in the early days Taylor and his group were held back by lack of funds. So was everyone else. We were in a depression. For instance, about 1937 the only work going on in the giant turbine shops of the General Electric Company in Schenectady was for the Navy. After the Naval Appropriations Committee of the House of Representatives had inspected the research work underway at the Naval Research Laboratory and had seen radar at work with their own eyes, there was never any trouble getting funds from Congress. In fact, Congress gave the Laboratory tremendous moral support in radar and everything else. The trouble was in getting past the Bureau of the Budget which I generally by-passed with the aid of the Appropriations Committee.

The development of radar was unique in the history of science and technology in that the development of theory, the design and manufacture of equipment and its installation, the training of personnel, and upkeep proceeded simultaneously. Our scientists at the Naval Research Laboratory were never able to pursue one single subject continuously. At first no one could make a high-power transmitting tube although many laboratories were working to develop better tubes. Some radio engineers predicted that there was no future for radar because high-output tubes could never be produced. Schools for training operators and maintenance men had to be established by our scientists who also had to go to the Navy Yards to train yard workmen to install radar sets properly.

The basic patent application on a radar system with transmitter, receiver and indicator, all synchronized for measuring distance, and with a short-time constant receiver for withstanding the main bang of the transmitter and resolving short pulse echoes from the transmitter signal at short range was filed jointly by Young and Page. This patent was issued in May 1951. Nine other radar patents which have been issued to Page cover a number of items involving common antenna working, presentation systems—including PPI, precision range systems, and so forth. One of these entitled "Radio Frequency Pulse Transmission" is a basic patent on radar pulse transmitters. Basic patent applications also held by Page include self-quenching transmitters, IFF, automatic tracking in angle and range, radar guidance of missiles, and many other radar functions totalling over 70 cases in all. In addition, a few important applications are in the name of A. A. Varela for pulse-forming networks and pulse stretching, L. R. Philpott for high-voltage modulation, range indicating circuits, and IFF, and A. H. Schooley for range-gating and precision range indication.

The post-war experience of the Naval Research Laboratory has clearly demonstrated the wisdom of providing the Laboratory in March 1942, as previously noted, with its own patent lawyer who was independent of the Office of the Judge Advocate General and who closely integrated his patent services with research activities. An outsider would comment that this should have been done ages ago. Correct, but even then it was almost too early for the Navy.

Of the hundreds of applications filed on radar and related inventions, 70 have been placed in interference with similar applications filed on behalf of adverse interests. Each of these interferences represents a situation where, had the Government application not been filed, the Patent Office would have issued a patent to others who would then be in a position to press a royalty claim against the Government. Of the 70 proceedings to determine priority of inventorship, 17 have gone to final judgment or have been settled. In only one of these cases has the Government lost the rights to use the invention.

Yet, in spite of this brilliant record, Dr. Louis N. Ridenour, in his book entitled *Radar System Engineering*,[19] states: "The research agencies of the American Army and Navy had a long and complicated history of early experiment, total failure, and qualified success in the field of radio detection." This seems to me to be a wholly irresponsible statement.

The truth of the matter is that if it hadn't been for the pioneer work in radar at the Naval Research Laboratory this country would have entered World War II without any radar and, when the British disclosed their limited accomplishments in radar to us, we would not have been able to understand what they were talking about. Finally, remember that the decisive battles of the Coral Sea and Midway were fought exclusively with only Naval Research Laboratory radar. Microwave sets, stemming from the British disclosure of the magnetron, arrived very slowly all through the remaining battles of 1942—Savo Island, Eastern Solomons, Cape Esperance, Naval Battle of Guadalcanal and Tassafaronga. Western Electric's 700 megacycle fire control radar figured in these battles as early as Savo Island, although improved models finally embodying the magnetron but still operating at 700 megacycles were conspicuous later.

In closing my remarks in regard to the Navy's connection with the invention and development of radar I would like, as I have done so often before, to pay my tribute to the subcommittee on Naval Appropriations of the House Appro-

[19] McGraw-Hill Book Company, Inc., 1947, page 14.

priations Committee. Funds were always available to me, whether it was for high-pressure high-temperature steam, diesel engines, or radar.

Atomic Energy

After announcing one day in April 1939 to the Secretary of the Navy's Council the possibilities of the use of atomic energy in bombs and in the powering of ships and being greeted by the usual audience of fish eyes, I was buttonholed by one Captain Gilbert C. Hoover of the Navy. Captain Hoover, it seems, was a member of a secret atomic energy committee under the direct surveillance of the President of the United States. It was all news to me. Dr. Lyman J. Briggs, then Director of the Bureau of Standards, was the Chairman of this committee. Captain Hoover was shocked that I was discussing, even in the Council of the Secretary of the Navy of the United States, such a hush-hush project as atomic energy. I hadn't the slightest suspicion that it was hush-hush.

You will remember that in the old days when Porfirio Diaz was President of Mexico, he had a very good system. Whenever a bandit became too troublesome, Diaz immediately incorporated said bandit into the Mexican Army. In accordance with the policy of Porfirio Diaz, I was incorporated into the President's super-secret atomic energy committee. I was succeeded by my good friend and fellow worker, Dr. Ross Gunn, of the Naval Research Laboratory and finally both of us, through successive reorganizations, were eliminated—a painful Washington process originated by politicians and admirably perfected by college professors, among others.

What happened was this: The first announcement of uranium fission was made in this country on 26 January 1939. On 20 March 1939 Dr. Ross Gunn of the Naval Research Laboratory came to my office and reported that nuclear fission had been accomplished and that he thought it offered outstanding possibilities in submarine propulsion. He requested me, as Chief of the Bureau of Engineering, to allot $1,500 to the Naval Research Laboratory to enable them to carry on preliminary exploration work. These funds were granted and work on the project was begun on that date. This $1,500 and succeeding funds were, to the best of my knowledge, the first government money spent on the study of atomic fission. This employment of government funds in connection with atomic energy antedates by almost a year "the first transfer of funds ($6,000)" recorded in paragraph 3.6 of the famous Smyth Report on the activities of Manhattan District. This work at the Naval Research Laboratory proceeded quite independently of the formation and operation of the committee appointed by President Roosevelt in connection with atomic fission and quite independently of my fleeting appearance on that committee.

I had announced at the meeting of the Council of the Secretary of the Navy the work that we were doing at the Naval Research Laboratory and the possible employment of atomic energy not only in the making of bombs but also as a source of power in driving ships. It was this latter phase which interested us because the horrible facts incident to the explosion of the atomic bomb were quite accurately anticipated at that time and no one of us ever expected that the United States would use what seemed to us then such an inhuman method of warfare.

The first step appeared at that time to be the production of large amounts of uranium-hexafluoride, and subsequent advances have shown this evaluation to be correct. This chemical compound was the simplest and most stable known form of uranium that occurred in the gaseous state under normal conditions. Its production on a large scale was indicated even then. Using the first $1,500 allotted by the Bureau of Engineering, Mr. R. R. Miller of the Chemistry Division, Naval Research Laboratory, working with Dr. T. D. O'Brien who was specially hired from the University of Maryland for this work, undertook the production of modest amounts of the key chemical. The first method employed metallic uranium which was pyrophoric and was unnecessarily dangerous and expensive in production. Small amounts were produced, however, and nine months later,

10 January 1940, the Laboratory supplied gram-size samples to the University of Virginia and the University of Minnesota for use in preliminary isotope separation work.

As early as 1 June 1939 the Laboratory had submitted a memorandum on the use of atomic power for driving submarines. In 1940, the Laboratory entered into contracts with the Carnegie Institution of Washington, the University of Virginia, and Columbia University for the purpose of getting a power program started involving the production of material suitable to operate a chain-reacting nuclear pile. Close liaison was maintained with Professor A. O. Nier of the University of Minnesota who made many of the early isotope measurements without payment. Thirteen thousand dollars was allotted to the University of Virginia to carry on the early phases of isotope separation by use of high-speed centrifuges. Dr. Jesse Beams was a world authority on such high speed devices, and through cooperation with him the Laboratory as early as February 1940 had a fraction of a gram of moderately enriched uranium 235 isotope. Simultaneously, $30,000 was assigned to Columbia University for the study of centrifugal fractionating columns, also for isotope separation. In addition to this, the Naval Research Laboratory bought, for the use of the Columbia group, one gram of radium which they mixed with beryllium in such a way as to provide a powerful neutron source. This powerful neutron source was of use to the Columbia group in their early study of pile performance using ordinary uranium and graphite.

In addition to the allotments mentioned above, $3,500 was finally assigned to the Carnegie Institution of Washington, to help support the liquid thermal diffusion method of isotope separation. The early costs of this investigation were borne wholly by the Carnegie Institution as a public service.

Beginning in January 1941, Dr. Philip H. Abelson of the Carnegie Institution started working for the Laboratory. Recognizing the inadequacy of the methods then employed for the production of uranium-hexafluoride, he succeeded in working out a greatly improved method which has been widely used. The production phases of this improved method were sufficiently worked out by 1 December 1941, so that large scale production was possible. This later method was disclosed to the Harshaw Chemical Works, 1 December 1941, was successfully employed on a large scale commercially, and was the basic material used in the Naval Research Laboratory isotope separation.

The Naval Research Laboratory very early recognized the practical importance of isotope separation and foresaw that it would be a necessary step before atomic power could be made available for naval use. Several methods for isotope separation were investigated or invented but showed little promise. However, in 1940, the work of Dr. Philip Abelson, then experimenting on liquid thermal diffusion processes at the Carnegie Institution of Washington, attracted the attention of Dr. Gunn. Because of the highly practical nature of Dr. Abelson's approach to this difficult problem, the Naval Research Laboratory first financed the work at the Carnegie Institution but, in July 1941, Abelson and the entire project were transferred permanently to the Naval Research Laboratory. The Abelson-Gunn method of isotope separation which resulted was a relatively simple and efficient method for enriching the active constituent of uranium from the natural value of 0.7 per cent up to any necessary value. Within a limited time it is usually not very practical to enrich the active material above 10 per cent. Pilot plants in succession were built and operated at the Naval Research Laboratory. One was a 10-15 column isotope separation plant, very nearly identical with the type finally adopted at Oak Ridge, and was operating by 15 November 1942. As a matter of fact, Dr. Gunn early envisioned the magnitude necessary for a plant with sufficient capacity to mass produce concentrated U-235, but I felt that it would be better judgment to build larger and larger pilot plants until we knew more about the engineering problems involved. As Secretary Forrestal so correctly said later, "We couldn't have won the war with men who were brought up in a depression economy; we needed men who could handle money with shovels."

Eventually, in 1943, a large plant of 300 full-size units suitable for full-scale production was erected at the Philadelphia Navy Yard where steam in sufficient quantities was available. This plant was so designed that concentrations approximating 0.9 per cent of active material could be produced directly and, by a simple modification of the interconnections, still higher concentrations could be produced. While the above increase in the concentration of the active constituent does not sound impressive, it was a logical choice for material which would be fed to a magnetic separator since such a plant is not adapted to handle large amounts of active material but does carry the material on to high concentrations. The output of highly concentrated material from the magnetic separation plant was increased 30 per cent by this use of the partly concentrated active material from the Navy's Philadelphia plant.

In December 1942, a couple of months before I left the Naval Research Laboratory to become a special assistant to Mr. Forrestal for plant seizure and scientific work not connected with atomic energy, a Colonel of the Army Engineers, Lester R. Groves (later General), came unheralded to my office at the Naval Research Laboratory and told me that he would like to see the pilot plants which we had built for the concentration of U-235. He gratuitously announced that he knew nothing at all about atomic energy and would be very glad to receive from us all the information we could give him. Such information was given him gladly. We always made complete disclosure to appropriate agencies such as the National Defense Research Committee and the Office of Scientific Research and Development. Groves, it appeared, had taken charge of Manhattan District in September 1942. After his visit and before the end of 1942 I was transferred to the Navy Department and thereafter had no connection with atomic energy.

The Laboratory was visited by a Manhattan District advisory committee on 23 January and 27 February 1942. While its technical report was favorable to the Naval Research Laboratory process, its utilization was vetoed by higher-ups and nothing further was done by the Manhattan District until after June 1944, which marks approximately the date of completion of the 300-column isotope separation plant at the Philadelphia Navy Yard referred to above.

Manhattan District realized in June 1944 that, if we were ever to have a bomb, use must be made of both the magnetic separation and the thermal diffusion methods. At that time General Groves visited the Naval Research Laboratory 300-column plant at Philadelphia. On 26 June 1944 the blueprints of the Naval Research Laboratory plant were turned over to Colonel Fox of the Manhattan District to facilitate the construction at Oak Ridge, Tennessee, of a new plant using the Abelson-Gunn process with 2,200 elements or seven times the capacity of the Navy plant at Philadelphia. The Naval Research Laboratory began training operators for this plant in December 1944. This new plant was completed early in 1945 in time to facilitate the final refining of U-235 by the magnetic method for use in the first atomic bomb. As a matter of fact, the Naval Research Laboratory proposed its method as an insurance against the failure of the Manhattan District's method as early as June 1943, but it was not until November 1943 that permission was obtained by the Naval Research Laboratory from Manhattan District to proceed on a limited basis. The Navy specifically agreed not to build up its research team to any appreciable extent.

In 1939 the Naval Research Laboratory was the only Governmental agency working on atomic energy. Presently outsiders became interested in the work and a special S1 Committee was set up in September 1940 with Dr. Lyman J. Briggs, Director of the Bureau of Standards, as Chairman. Dr. Gunn joined this committee 17 September 1940. The committee functioned well, with a free interchange of ideas and material facts, until early in 1941 when the work of the committee attracted the attention of a number of members of the National Academy of Science. This new group moved in on the original committee of active workers and evidently took charge because, by April 1942, they had issued orders to "hold reports . . . without distribution to members of the Section S1." The effect of this order was to deprive the Naval Research Laboratory of all information regarding developments in atomic energy. After Major General Groves took charge of Manhattan Project in September 1942, the isolation of the Laboratory became complete. It was rumored that complete information was turned over to certain British scientists. Possibly Claus Fuchs was among them, but as far as the Naval Research Laboratory was concerned Manhattan District pulled down the iron curtain on the Laboratory—for what reasons, I do not know. Of course, the fault may have been in the Navy's high command because, at one stage, after the news got around Washington about the huge disbursements of Manhattan District, there were many who thought that the whole thing would be a flop, were afraid of the day of reckoning, and didn't want to be involved in it. Whatever the reason, the Naval Research Laboratory which contained Drs. Gunn and Abelson, the first two people in the Government service to work on atomic energy, was debarred from participation in this great and successful effort except for the one instance cited.

As an illustration of the lack of cooperation between the Laboratory and the Manhattan Project, one may note that the isotope separation development of the Naval Research Laboratory (that finally adopted by Manhattan Project after its original plans went awry) was held up in November 1943 by a Manhattan directive to the War Production Board which prohibited it from letting the Naval Research Laboratory have any of the key chemical which the Laboratory itself had developed. Upon the decision of the Laboratory to make the chemical itself, the Laboratory was not permitted to purchase the raw material which was in plentiful supply. This particular situation was corrected only after great effort and not until June 1944. Meanwhile there was nothing going on in the Navy in regard to the atomic powering of submarines due to the isolation of the Naval Research Laboratory.

Another contribution I made, and I do not find any reference to this in Dr. Smyth's otherwise excellent presentation—furthermore, this will give you an idea how tight money was in 1939—was a passing of the hat around the various Bureaus of the Navy Department and the Ordnance Department of the Army to obtain enough money (about $20,000) to purchase a gram of radium which we promptly loaned to Columbia University and which Columbia just as promptly overheated and scattered through the ventilating ducts of their laboratory. Columbia later used this radium, after its recovery, in their studies.

Looking back over the problems faced by the Naval Research Laboratory in their development of atomic energy, many of us were dismayed by the arbitrary decisions and lack of interest in the public welfare in high places. I daresay that the military possibilities of submarine navigation under the polar ice caps are still not appreciated, although they were apparent even in 1939 to those pushing the Navy atomic energy program. The isolation of the Navy from the main program and the political chicanery that even the Secretary of the Navy could not correct were indefensible in time of war and delayed the arrival of the atomic bomb by many months.

Light Armor and Plastic Armor

Because of the Nye Committee hearings and the attempt by some to pin the responsibility for World War I on the "merchants of death," there was no collaboration between the Navy and the powder and steel manufacturers prior to World War II. So much for the loudmouths.

An incident will highlight this situation.

When World War II broke out in Europe, aviation circles were shocked to learn that the Germans were using light armor in planes. Who knew anything about light armor? Why the Naval Research Laboratory of course. The Laboratory for years, under the leadership of Dr. Ross Gunn, had been trying to do two things at once: First, to discover the laws, if any, which controlled the phenomenon of penetration; and, second, to ascertain whether or not any law of similitude existed between projectile penetration of light armor and heavy armor. Every promising substance or combination was tested for use as light armor. One day I saw a magazine picture showing Henry Ford using a sledge on an automobile fender made of plastic. We tried plastic.

Two things came out of this experimentation at once which were totally unexpected. First, the Laboratory was discovered to be the country's most experienced source on the behavior and characteristics of light armor; and, second, the work on plastics sparked further and successful work on bullet-proof vests. That's research—while looking for one thing, something else bobs up and often a much more important discovery than the one originally sought.

Precipitation Static

For years Dr. Gunn, even back in the days of the Navy's dirigibles, had worked on aircraft problems arising from unusual atmospheric electrical conditions. Under certain meteorological conditions, raindrops or snowflakes will transfer electrical charges to a plane when they strike it. The resultant static charge on the plane will build up to more than a half-million volts or until it relieves itself by corona discharge. Since this discharge is pulsating and broadcasts on all frequencies, it is obvious that radio communication with or to the plane is out when it flies in these unfavorable conditions. No one, Army or Navy, had been interested in these hazardous conditions resulting from so-called precipitation static. The British said it did not occur in their latitudes. Nevertheless, one day during the war an Army plane crashed in foggy Iceland, killing all on board including a General. Corona discharge probably rendered its radio receiver inoperative, and orders to return to point of departure were presumably never received. This accident sparked official interest in the serious problems of precipitation static.

Dr. Gunn had been doing a lot of flying between Minneapolis and Fairbanks, Alaska, experimenting with precipitation static. He already had it under some control by trailing from the plane a device which allowed the static charge to leak off to some extent.

Now everyone became interested in the subject, business boomed, money poured in, and ample facilities were supplied at Minneapolis to further Gunn's work. This support resulted in a quite satisfactory solution to the problem that is now practiced in all long-distance flights. Even the poor General got into the act—they named an airfield after him.

Hydrogen Peroxide Fuel

The Naval Research Laboratory, again under Dr. Gunn, had experimented for a long time with concentrated hydrogen peroxide as a source of oxygen for the diesel engines of submarines running submerged and for use in torpedoes. As a matter of fact, a long-range hydrogen peroxide torpedo that ran quite wakeless was developed by the Naval Research Laboratory and range tested at Newport, Rhode Island, during September 1937, but there was not enough interest to develop it beyond the laboratory stage until the German achievements in this respect became available *after* the war. As early as 1935, Dr. Canfield, who was also identified with the hydrogen peroxide torpedo, tried unsuccessfully to get the Bureau of Aeronautics to support a rocket development program to cost $5,000. We could have had submarines with high underwater speed and schnorkels long before the Germans if anyone in the high command had asked for them and if the General Board had insisted on them as military characteristics.

Development of Potassium Superoxide as an Oxygen Source[20]

While I was Director of the Naval Research Laboratory, some of the most spectacular developments took place in the solid oxygen problem. Early in the 1930's, Dr. Borgstrom, Head of the Chemistry Division, and Mr. Hobson,

[20] The assistance of Mr. R. R. Miller, Head of the Physical and Inorganic Branch, Chemistry Division, Naval Research Laboratory, in preparing the story of potassium superoxide is gratefully acknowledged.

of the Bureau of Construction and Repair, had planned research work on oxides containing more available oxygen than sodium peroxide which was commercially available. Attempts were being made to use sodium peroxide with a catalyst, but they were unsatisfactory in the apparatus desired by the Bureau of Construction and Repair since the oxygen furnished from the reaction with CO_2 gave less than half as much oxygen as is required in a breathing cycle. Normally, in breathing, $\frac{8}{10}$ of a volume of CO_2 is expired for each volume of oxygen consumed which means that in a rebreathing cycle using an oxygen generating material the reaction of a CO_2 volume should give approximately 1.2 volumes of oxygen. Sodium peroxide gave ½ volume of oxygen per volume of CO_2 reacting. The compound in which the Laboratory was interested was potassium superoxide which gave a volume and a half of oxygen for each volume of CO_2 reacting. Research work in this field was started in 1936, and it was determined that potassium superoxide was suitable for the reaction with water vapor and CO_2 and could be used in the breathing cycle with an excess of oxygen available.

In the fall of 1939 it was shown that the quantity of potassium available was not sufficient to be used for the production of this oxide and furthermore that a mixture of sodium and potassium oxides could be made from the oxidation of an alloy of alkali metals which was suitable for a rebreather. The mixed oxide was not of as high an efficiency as the potassium superoxide but was sufficient to furnish a slight excess of oxygen in breathing canisters. It could be made by the treating of potassium chloride with sodium metal in a proportion of 25 pounds of salt to 5 pounds of sodium metal with the vaporization and burning of resulting alloy. A large scale laboratory examination of this process was started in the fall of 1939 and led to the formation of a pilot plant which would produce 50 or 100 pounds per fourteen-hour day. This material was adopted as a satisfactory oxide for the rebreather canister of the Navy oxygen breathing apparatus. Early in 1941 the pilot plant was enlarged to three units which were operated twenty-four hours a day, seven days a week, and in the course of a year this plant produced approximately 60,000 pounds of the mixed oxides for filling Navy canisters. At the end of the year's operation, the process was shifted to industry which since that time has furnished the Navy oxygen rebreather material.

The shipboard use of rebreathers containing this oxide was responsible for the saving of numerous lives during World War II.

This same process of burning was copied and used by the DuPont Electro-Chemicals Division in the production of potassium superoxide which was shipped to Russia on lend-lease during the war. During the operation of the plant and afterwards, research was done at the Naval Research Laboratory on the process of producing potassium for a superoxide which led indirectly to the development of a commercial process which now produces potassium for filling Navy canisters with the superoxide.

Developments were carried out to adapt the superoxide or mixed oxide to high altitude flying apparatus and, while results appeared encouraging, the plans were dropped with the adoption of the Demand oxygen gas regulator which has been in general service use since its development.

Sonar—Sound Navigation and Ranging

No chapter on the Naval Research Laboratory would be complete without reference to the development of underwater sound. For years Dr. Harvey C. Hayes, the head of this division, hung onto this problem with the tenacity of a bulldog, and it was my pleasant duty to back Dr. Hayes when others had failed to appreciate his brilliance and high sense of purpose.

The Sound Division of the Naval Research Laboratory started as an Experimental Station established by the Navy at New London in 1917 and closed in 1919. The mission of the station was to devise and develop ways and means of detecting and locating a submerged submarine. Preliminary tests confirmed a prediction from theoretical considerations that the most promising approach to a solution of the problem lay in the field of submarine acoustics.

The efforts of the station during the two years of its existence were largely devoted to the development of submarine sound receivers. The name "hydrophone" was applied to these devices. Of the numerous types that were developed, none met the requirements for directing a submarine attack. Either they lacked the necessary directivity or else they could not operate while underway because of excessive local noise background.

Thus, by the end of World War I, the U-Boat had demonstrated its potentialities as a powerful future war weapon but, in the meantime, the Navy had developed no effective countermeasures. It was therefore decided to continue the antisubmarine work and, in particular, to determine the possibilities of the so-called MV Hydrophone which gave promise of meeting minimum requirements but had not yet taken final form. The MV Hydrophone was equipped with multiple variable sound receivers. The detectors were arranged in two multiple groups of twelve units each, installed on the port and starboard sides, fore and aft. The work program with reduced personnel was transferred to the U.S. Naval Engineering Experiment Station at Annapolis, Maryland, pending completion of the Naval Research Laboratory at Anacostia, D.C.

Research during the three-and-a-half year period at Annapolis was confined to frequencies within the audible range. The Sonic Depth Finder was a product of this period. Development of the MV Hydrophone was completed. Extensive field tests showed this device to be superior to any of its predecessors by giving fair ranges and directivity at speeds up to about 6 knots. At higher speeds and in a moderately rough seaway, when installed in a sub-chaser, water noise masked the propeller sound of the target. When this device was installed on a submarine, noticeably better ranges were obtained because of the lower noise background.

As a matter of interest, it may be stated that the principle of operation of the MV Hydrophone whereby the responses from a multiplicity of spaced hydrophone units are brought into consonance by electrical retardation lines forms the basis of the Sonic Hydrophone on the U-Boats of World War II.

The work at Annapolis led to the conclusion that the anti-submarine problem could not be solved by "listening devices" alone since they do not readily give the target range, fail entirely when the target rests on the sea bottom, and in general tend to strengthen rather than to weaken the target submarine because of the more favorable listening conditions on a submarine.

Development of ways and means for detecting and locating a submarine by means of signal echoes reflected from its hull was started in 1923 when the work was transferred to the Naval Research Laboratory. This resulted in the development of the XL ultra-sonic transducer[21] which, following the principle devised by Langevin, employed a sandwich type of projector consisting of quartz crystal slabs cemented between steel plates. It was sharply resonant at one frequency and was used in the range of 25-40 kilocycles.

Exhaustive tests of this equipment in 1928 gave reliable results to ranges approximating 1,200 yards. Such ranges obviously were inadequate for purposes of search, but the new equipment greatly strengthened the attack once sound contact was made with the target. While the echo range could doubtless be increased through improvements suggested by the tests, it appeared improbable that they could be extended adequately to meet search requirements.

It was therefore decided to tackle the problem of extending the search range by listening to the ultra-sonic components of the submarine's propeller sounds. This led to development of the JK ultra-sonic receiver. The JK ultra-sonic receiver used Rochelle salt crystals as the sensitive elements. This transducer operated as a receiver in the frequency range of 10-50 kilocycles. The high directivity of this searchlight-beam type of receiver reduced the local noise background

[21] Transducer: "A device actuated by power from one system and supplying power in the same or in any other form to a second system. Example: telephone receiver."—Webster's Dictionary

sufficiently to increase the search range to roughly 5,000 yards while underway at speeds up to about 9 knots. The combination XL and JK, perfected by 1932, represented a distinct advance in antisubmarine equipment since the JK projector could be used during search operations to discover and bring a target within range of the XL receiver which then took over and directed the attack.

The use of Rochelle salt crystals as a weak ultra-sonic sound source during the study and development of the JK receiver encouraged a belief that such crystals could serve as an ultra-sonic projector of sufficient power for depth sounding purposes. This proved true and resulted in the ultra-sonic depth finder which in various forms is installed in practically all ships of the Navy.

The success achieved by use of Rochelle salt crystals, both for generating depth-sounding signals and for receiving their echoes reflected from the sea bottom, gave rise to a belief that the functions of both the XL transmitter and the JK receiver could be combined into a single equipment through further development of the Rochelle salt projector. This led to development of the QB equipment which gave roughly twice the echo range of the XL and, at the same time, equalled the listening range of the JK. Thus the QB was a Rochelle salt echo-ranging device. In construction it was identical with the JK except that the QB was designed to transmit as well as receive ultra-sonic energy. This QB equipment constituted another marked antisubmarine improvement.

Field tests of the original QB equipment during the summer of 1934 corroborated predictions that the new projector could operate effectively over a relatively wide frequency band (17-30 kilocycles) and permit each of several cooperating patrol ships to tune their equipment to a different frequency and thus minimize interference. They also proved, as predicted, that the lower frequencies gave improved search ranges and that the higher frequencies, with the resulting narrower sound beam, served best for directing an attack. Finally, the tests disclosed the need for a tuning-control system that would insure like tuning for both driver (projector) and receiver-amplifier if full advantage were to be taken of the wide frequency range of the QB projector.

Recognition of this need led to the development of the so-called "uni-control system" whereby the driver and inverted superheterodyne receiver were coupled and tuned by a single inter-coupled variable frequency oscillator to like frequency throughout the frequency range of the projector. This arrangement permitted a fixed intermediate frequency for the receiver-amplifier and thus allowed the use of band-pass filters in the intermediate stage that compensated at all frequencies for loss of selectivity due to the low Q value of the QB projector.[22]

Development of the uni-control system was completed in 1939. The QB type projector in combination with the uni-control driver-receiver system represented another marked improvement in antisubmarine equipment. This combination played a stellar role in the Battle of the Atlantic.

The variation and lack of reliability of echo ranges obtained with the QC and early QB systems indicated that the physical characteristics of the medium (sea water) were less stable than was anticipated, and that the echo ranges in general fell far short of those predicted from the classical theory of absorption based solely on the viscosity of the medium.

The QC was an echo-ranging underwater sound equipment. Its projector was of the magnetostriction tube-and-plate type based on the early work and patents of Professor G. W. Pierce. In it a large number of nickel tubes, each 3″ long and ⅜″ in diameter were embedded (force fit) in one side of a common circular steel plate. The opposite side served as a radiating face and was exposed to the water. A coil was wound on each nickel tube and carried both the direct current

[22] The low Q value of the QB projector indicates that it was useful for sound generation or reception in water at any frequency within a relatively broad band, this band being centered on the fundamental frequency of longitudinal resonance in the projector's mechanical structure. Thus Q value specifies sharpness of resonance or frequency selectivity.

for polarizing the nickel and the alternating signal current which forced the tube-and-plate structure through the magnetostriction coupling to vibrate at its fundamental longitudinal resonance. Under excitation the radiating face moved in a reciprocating-piston-like manner and had sufficient area to develop directional radiation. This type of transducer is still in existence on many of our ships and was used considerably during World War II in the larger antisubmarine ships such as destroyers and light cruisers.

To account for the discrepancies in echo ranging noted above, a study of the propagation of sound in sea water was started in 1936. This work led to the conclusion that downward bending of the sound signal path caused by vertical temperature gradients was largely responsible for the non-uniformity of echo ranges at various times in the same locality or between different localities. This study also disclosed the existence of temperature inversions, predicted their screening effect on a submarine beneath such layers, and outlined a procedure for estimating the echo range as a function of temperature gradient and target depth. This information worked to improve the escape tactics of our submarines and our attack procedure against the U-Boats.

The development of a projector that would generate a supersonic signal of maximum intensity or, in other words, that would drive the sound-generating surface to cavitation amplitudes, was also undertaken in 1936. The so-called Electrodynamic (E.D.) Projector, built on the principle of a loud-speaker, which accomplished this purpose, was completed in 1939. By this time, however, the researches in propagation indicated that under many, if not indeed most, sea conditions encountered in practice no worthwhile increase in echo ranges should result from increasing the signal strength beyond that given by the QB projector.

This prediction was supported in part by comparison tests which showed that the more powerful E.D. signals gave no marked increase in echo range, but that at the limiting ranges they did decrease the percentage of signals that failed to return echoes. Wherefore it was concluded that the advantages gained by the use of powerful signals were not sufficient to warrant the increased weight and volume of the equipment and that such gains as the more powerful signals provided should be sought through reduction of noise background.

Following this decision an uncompleted undertaking of surrounding the projector with a sound-transparent streamlined form that was started four years earlier was revived. The earlier work had indicated that a form fabricated of thin steel with an interior reinforcing frame should provide the necessary mechanical strength and permit transit of the ultrasonic signals and echoes with no great loss of intensity. Development of such a "dome shield" of simple design that telescoped over the standard projector and was mounted by welding to the hull was completed by December of 1940. Field tests proved that the dome shield reduced the noise background to a point where both listening and echo ranging became practical up to speeds in excess of 15 knots. These favorable results led to a hurried dome installation program that eventually equipped every patrol ship with a dome-shielded projector and thereby greatly improved their effectiveness against the U-Boats.

The dome-shielded QB projector with uni-control driver and receiver-amplifier must be rated as one of the most powerful antisubmarine developments. Moreover, the same or similar equipment without the dome shield has been the main, and in many cases the only, sound equipment on our submarines. The QB submarine equipment has proved to be a powerful aid during search and evasion procedures, and instances are not lacking where a torpedo attack has been carried out with the aid of sound range and bearing data alone without exposing the periscope.

Such, in brief, is the history of the development of Sonar during the period intervening between the two world wars. The beginning of this period found the Navy sadly lacking in ways and means for combating the relatively simple U-Boats of that time. The end of this period found the Navy with completed developments in the field of submarine acoustics that served to defeat the tougher and more versatile

U-Boats of World War II and to increase the effectiveness of its own submarines.

These antisubmarine developments of the Sound Division of the Naval Research Laboratory contributed in no small measure toward bringing warfare against the U-Boats to a status that elicited from Grand Admiral Karl Doenitz a recovered order that states in part:

> "For some months past, the enemy has rendered the U-Boat ineffective. He has achieved this object, not through superior tactics or strategy, but through his superiority in the field of science; this finds its expression in the modern battle weapon—detection. By this means he has torn our sole offensive weapon in the war against the Anglo-Saxons from our hands. It is essential to victory that we make good our scientific disparity and thereby restore to the U-Boat its fighting qualities."

General Sound Research, 1918-1945

The need for ways and means of making underwater sound measurements in absolute units was recognized even before the work at New London was started, but the difficulties to be overcome and the time and manpower required to perfect such means delayed the undertaking until about 1934. By that time it became apparent that such measurements were required to determine the efficiency of the ultra-sonic projectors under development and that the ability to make such measurements would permit formulating definite performance specifications for projectors in terms of sound power output.

The problem was undertaken in this same year, 1934, and by 1938 this pioneering program had resulted in the development of both primary and secondary standards that have helped to coordinate and properly to evaluate the far-flung projector developmental program of the war effort.

The sound-detecting equipment developed and at hand at the beginning of the war could locate a U-Boat to within a range of roughly 10 yards and bearing to within plus or minus one degree so long as the sound beam could be held on the target. But as early as 1934 the modern U-Boat proved capable of submerging to depths below a horizontally directed sound beam at the shorter ranges involved in an attack. The early-visioned need for a projector mount that would permit measurably tilting the sound beam downward was thus proved to be real. Modification of the standard QC and QB projector mount to permit the sound operator to tilt the projector measurably was completed and thoroughly tested by late 1942.

Analysis of the probable tactics of U-Boats operating in pairs or groups indicated the possible hurried need for two independent sound equipments on antisubmarine patrol ships, one for carrying out an attack and the other to stand guard against being attacked by a second or cooperating U-Boat. The solution of this problem resulted in a dual installation wherein the two projectors, mounted one above the other on concentric shafts, could be trained independently about the azimuth and, in addition, the upper one could also be tilted to maintain short range contact. The standard QC and QB equipment could be replaced by the dual system using the same hull fittings, the same hoist and train mechanism, and the same dome. This development was completed and successfully tested early in 1943.

Two remote-controlled hydrophone systems termed "Herald," a simple one of medium range and a more powerful streamlined design for use in locations subject to strong tidal currents and resulting high noise background, were completed in 1943. The term "Herald" is an abbreviation for Harbor Echo Ranging and Listening Device. It was a bottom-mounted Rochelle salt transducer employed in guarding the harbors and advance bases against submarines and sneak craft. Also, in the same year, the Sono-Radio Buoy was completed in cooperation with the Radio Division, as previously noted. These devices found wide use in protecting harbors and fleet bases.

There was also developed portable equipment for locating

leaky valves in aircraft hydraulic control systems through reception of the sound generated from such leaks.

Technique was developed for determining the acoustical performance of an ultra-sonic projector by tests made in a tank through the use of short impulse type of signals that could be received and recorded before the arrival of distorting echoes from the sides of the tank.

The end of hostilities found the Sound Division with two major programs of applied research that were well under way. One concerned the development of a gyro-stabilized capsule-enclosed projector for antisubmarine patrol ships, and the other an ambitious pro-submarine program calling for a dome-enclosed remote-controlled hydrophone embodying many of the principles employed in the harbor defense hydrophone and for methods and means of indicating on a cathode ray screen the bearing and range of every object in the projector beam that returns an echo from a single intense impulse type of signal.[23]

When we traded information with the British in 1940 it was interesting to note that we both had made practically the same progress in underwater sound. There were some differences, however. Our equipment, sound or any other, was better made than the British and had already been developed for mass production. The second great difference was that the British had developed a splendid school or method for training underwater sound operators and we had nothing. Thus they had made the original contribution of synthetic training for operators. Our Office of Fleet Training had concentrated on blaming the equipment for *all* failures to obtain satisfactory results in tactical exercises and had made no contributions to training at all. In the Royal Navy, when something goes wrong, they seem to blame the personnel, whereas in our Navy we blame the material.

The art of synthetic training is not new. The Germans developed it highly before World War I to train submarine

[23] Much of the information on Sonar is derived from "A Brief History of the Sound Division, Naval Research Laboratory," by Harvey C. Hayes, former Head of that Division.

commanders. We copied and used that method. The synthetic training of airplane pilots was well and early developed in this country before World War II, and important contributions were made to the art when the Special Devices Division was under the command of Rear Admiral Luis deFlorez, USNR. We missed out on underwater sound training, I suppose, because of the lack of vision and the lack of appreciation of the universal application of this type of instruction.

This British contribution to the art of undersea warfare is but a slight indication of the astounding amount of improvisation they accomplished during the Battle of Britain, particularly in its early stages when they were so woefully unprepared.

While I was Director of the Naval Research Laboratory, I was fortunate in having so many able commissioned assistants. My first Executive Officer was Captain (now Vice Admiral) Robert P. Briscoe, an exceptionally able and competent assistant who was all over the place and always looking. His enthusiastic support of all my attempts to improve the Laboratory and the Navy, together with his understanding interest in research and development problems, were invaluable. He was relieved by Commander Robert J. Walker, retired but called back to active duty. Walker had been in industrial life and brought useful information to us about "how they did it on the outside." He was particularly good on organization, and when I went with Mr. Forrestal I took Walker along to organize my new office. He proved his dependability when I was away from this office on long assignments.

Commanders (now Captains) Martin J. Lawrence and Joseph R. Garvin were, on different occasions, shop superintendents. Lawrence was Executive Officer when I took him from the Laboratory to help me start the Office of Naval Research and to become its first Executive Officer. While Garvin was shop superintendent, we drew up plans for the first new laboratory buildings to be built in years. Through my friendship with the late Dr. Frank W. Jewett, then Director of Bell Telephone Laboratories, I was able to utilize in the plans for the new buildings all the latest information which Bell Laboratories had acquired from Europe and elsewhere and which they were incorporating in their new laboratory then under construction at Murray Hill, New Jersey.

When radar began to get too big to handle, I talked Commander (now Captain, retired) David R. Hull, who was a post-graduate in radio and then Commanding the Bureau of Engineering's Experimental Radio and Sound Ship, into joining the EDO's and coming to the Laboratory to form and head up a front office radar section. Commander (now Captain) J. F. Mullen, an aeronautical engineer and pilot, came along later and was of the greatest help in airborne radar. Commanders (now Captains) A. F. Junker and W. B. Goulett were among others in a very remarkable and easy-to-work-with team.

5

Plant Seizure and Operation

AT VARIOUS TIMES, I seized and operated plants under Executive Order of the President of the United States as follows:

Federal Shipbuilding & Dry Dock Company, Kearny, New Jersey. 25 August 1941 to 6 January 1942
General Cable Corporation, Bayonne Plant, Bayonne, New Jersey. 14 August 1942 to 20 August 1942
Howarth Pivoted Bearings Company, Holmesburg, Philadelphia, Pennsylvania. 15 June 1943 to 30 June 1944
Remington-Rand, Inc., "N" Division, Elmira, New York. 29 November 1943 to 3 July 1944
Los Angeles Shipbuilding & Dry Dock Corporation, San Pedro, California. 8 December 1943 to 10 July 1944
Jenkins Brothers, Incorporated, Bridgeport, Connecticut. 14 April 1944 to 15 June 1944
104 Uptown Machine Shops, San Francisco, California. 14 August 1944 to 10 April 1945
Lord Manufacturing Company, Erie, Pennsylvania. 25 October 1944 to 10 April 1945

On one occasion I was operating plants simultaneously at Bridgeport, Connecticut; Elmira, New York; Holmesburg, Pennsylvania; and San Pedro, California. On another occasion I was operating plants simultaneously at Erie, Pennsylvania, and San Francisco, California. This wide coverage obviously would not have been possible without excellent assistants.

In addition to my plant seizure and operation work, I surveyed for Mr. Forrestal the Atha Works of Crucible Steel at Harrison, New Jersey; served as consultant to the plant of Triumph Explosives, Inc., at Elkton, Maryland, which the Bureau of Ordnance was operating after seizure; surveyed the main plant of Allis-Chalmers at Milwaukee, Wisconsin; and surveyed for Assistant Secretary Bard the New York (Brooklyn) Navy Yard and Brewster Aeronautical Corporation, Long Island City, New York.

Crucible Steel's trouble was insufficient government production. As a result of my survey, the situation was promptly remedied and Crucible won a Navy "E."

Allis-Chalmers, like Crucible Steel, wasn't producing enough. Everybody was to blame, company and government. The management had too much deadwood and did not show enough recognition of talent among the younger members. Office of Production Management was changing priorities so fast that an estimated $1,000,000 worth of partly-fabricated material was messing up the machine shop floors, and some of the Navy Bureaus were also complicating the situation. This survey also resulted in marked production increase.

Later, when Allis-Chalmers was strikebound, I was ordered to seize the plant and break the strike, using troops from Fort Sheridan if necessary. I was all set to leave Washington for Milwaukee one afternoon when the strike was called off the forenoon of the same day. Possibly the news of the impending seizure "leaked" and the local labor leader decided not to tangle with the military.

My survey of the New York (Brooklyn) Navy Yard indicated 10,000 unnecessary employees. When I turned in my report to Assistant Secretary Bard, he asked me "what shall I do with this report?" I replied, "Neither you nor anyone else will do anything with it. It is a hot potato." I was right; nothing was done.

My survey of Brewster Aeronautical Corporation did not reveal anything satisfactory. The type of building construction was not suitable for rapid and economical construction of airplane elements which were shipped to a branch plant in Pennsylvania for assembly. Labor was unruly and difficult to handle. Inventory was not up to date, and material was scattered in perhaps 20-odd storage places. Navy accountants believed that a careful inventory would reveal that the company was insolvent. I was sounded out on taking the presidency at $25,000 per annum. I couldn't see it. Somebody, probably the Government, talked Henry J. Kaiser into taking the job. I noticed he didn't stay long.

Twice during my operations I was authorized orally to use troops if necessary. I was shocked that such authorizations were not reduced to writing. Perhaps I should not have been shocked in view of the circumstances.

At one time I spent 24 hours in Forth Worth, Texas, straightening out a plant which was a principal source of badly needed 40 mm. shells. A local bank had suspended a loan which had been made to the shell manufacturer. A V-loan from the Federal Reserve was also involved. Criminal charges had been made involving the manufacturer and government inspectors. I arrived at noon and in the afternoon obtained from the manufacturer a signed, but undated, letter of resignation as President of his company. The next morning at a special meeting of the Board of Directors of the bank, I talked them into reinstating the loan and filing the undated resignation until it should be needed. Before the departure of my train I had time to inspect a very modern airplane factory with Mr. Amon Carter of Fort Worth and Mr. Buck, his lawyer, and to leave town with one of Mr. Carter's famous 10-gallon hats. The manufacturer was acquitted, as I told him he would be, by a Texas jury; and the flow of badly needed 40 mm. shells to the front was never halted for a second.

All of these operations were minor compared with the big ones carried out under Executive Orders.

Seizure and Operation of the Federal Shipbuilding and Dry Dock Company, Kearny, New Jersey

The plant of the Federal Shipbuilding and Dry Dock Company at Kearny, New Jersey, had been closed down because of a strike. Over the weekend of 16 and 17 August 1941, seizure or, as it was called in those days of cautious procedure, "takeover" was considered imminent. It was generally accepted that the Navy would have to support the Mediation Board by adopting in some form the Union Maintenance Clause. At the same time it was the opinion of the Navy Department that the Navy would be on firm ground only if it took over the plant in order to restore production and that, if the Navy would adopt a Union Maintenance Clause at Kearny, such action would be a departure from its long standing policy of not taking sides with either labor or management.

On Saturday morning, 23 August 1941, word was finally received from the White House that settlement of the strike was impossible and that the plant would have to be seized by the Government.

In view of this policy, there had been long and varied discussions in which many people were involved as to how the plant could be operated—would it be operated as a Navy Yard, should a corporation be set up to effect its operation? The net result was that it was finally decided to seize and operate the plant under an Executive Order.

Considerable confusion surrounded the taking over of this plant. The President's press release and Executive Order were received from Press Relations over the ticker. Contrary to expectations, the press release failed to state whether or not the plant would be operated by the Navy Department in accordance with the recommendations of the Mediation Board.

Federal Ship had already offered the plant to the Government for sale or for lease and, when they received the information with the signature of the President of the United States directing the Navy to seize and operate the plant, they were as much in the dark as to procedure as anyone else.

At six o'clock that Saturday afternoon a meeting was held in the office of Mr. Adlai E. Stevenson, Special Assistant to the Secretary of the Navy. Present also were Messrs. John Vincent, W. John Kenney—Counsel to Mr. Forrestal, Norwood P. Cassidy—Special Assistant to the Chief of the Bureau of Supplies and Accounts of the Navy Department, and myself. It was agreed that the position on the labor question should be that the instructions from the Secretary of the Navy to me would be to observe the wage rates, hours, and working conditions prevailing immediately prior to the stoppage of work at the plant, subject to such modification as had been agreed upon; that having no instructions on the Mediation Board question, it was decided to make no commitments in that respect. There was some discussion as to what should be done if the union refused to withdraw the picket line, but the consensus of opinion was that this refusal was unlikely.

I had been designated to seize and operate the plant, and my party and I arrived at Newark shortly after noon on Sunday, 24 August. We were without Executive Order or written instructions of any kind.

My staff and I had discussions about crossing the picket line. In fact, we had had these discussions all the way up from Washington, anticipating as we did trouble with the strikers and expecting no trouble with management. Attempts were made informally to have the pickets withdrawn before our entrance to the plant of Federal Ship, but there was not sufficient time to make the necessary contacts, and anyhow I had decided that it was one of those cases which called for rapid action.

With a station wagon borrowed from a local Naval activity and with the American flag flying, we went through the picket lines of the Kearny Plant at a goodly rate of speed and with all the strikers cheering and waving American flags. Quite a different reception than we had anticipated. As will appear later, the union felt as if it had won a victory over management.

The seizure party arrived at the plant at 2:30 p.m. in accordance with an appointment made by Mr. Ralph Bard, Assistant Secretary of the Navy, with Mr. Lynn Korndorff, President of the Federal Shipbuilding and Dry Dock Company. The party was taken to Mr. Hemingway's office (the Vice President and General Manager), was told that Mr. Korndorff was temporarily engaged and would see the Admiral in a few minutes. After a delay of ten or fifteen minutes, word was received that Mr. Korndorff was free, and the party adjourned to his office.

Mr. Korndorff's attitude indicated doubt on his part as to whether the United States Steel Corporation, of which the shipyard was a subsidiary, would permit the Navy Department to take possession of the plant. He appeared still to be under the impression that it was a question for negotiation and that consent of the company as to financial terms, etc., was necessary.

A long and profitless conversation took place regarding all of the reasons why the Navy could not open up the Yard until certain conditions were met. The purpose of all this conversation was obviously to gain time and to delay as long as possible the seizure, reopening, and operation of the plant.

At about the time that negotiations were stalled in respect to when the plant could be opened, the Under Secretary of the Navy, Mr. James Forrestal, arrived and joined the conference in the office of Mr. Korndorff. While Mr. Korndorff changed his attitude to a certain degree after the appearance of the Under Secretary, he still was unwilling to commit himself without seeing his lawyers. In justice to him, it must be said that he had apparently been making an attempt to do just that since Saturday. He complained about the inability to get a copy of the Executive Order; he had to read it in the *New York Times*. He was told that, in this respect, he was in no different position than the Navy Staff, since I had not yet received an official copy from the White House. It was pointed out to him that there was nothing his lawyers could tell him which would affect the immediate situation unless, of course, they wanted to advise an injunction.

I had stated that I was most anxious to open up the plant Monday morning in line with the expressed expectations of the President in his press release. Mr. Korndorff scouted this idea, saying that it would be impossible to organize work for the men, etc.

The question of unloading freight that had accumulated

in the yard during the strike was discussed at length, as well as the difficulty of getting the supervisory force ready for opening. The problem was made to appear one which might take over a week to solve. Finally it was agreed that the plant would open full force with the first shift, Tuesday, 26 August, and that on Monday only the superintendents, department heads, foremen, assistant foremen, and sub-foremen would be called in.

The following notice was agreed to by the company and released:

> "First steps toward reopening the plant of the Federal Shipbuilding and Dry Dock Company will be taken at 8 a.m., Monday, August 25, when superintendents, department heads, foremen, assistant foremen and subforemen will report to complete plans for operations. Further steps in the resumption of work at this plant will be announced subsequently in the press and over the radio. It it expected that this plant will resume full operation beginning with the day shift at 7:30 a.m., Tuesday, August 26, 1941."

At about 3:45 p.m., Mr. John Green, President of the Industrial Union of Marine and Shipbuilding Workers of America, arrived at the Yard and was introduced to me. Mr. Green stated that he wished to tender his cooperation to the Navy with respect to the opening of the plant. I thanked Mr. Green and assured him that fair and equitable treatment would always be afforded to the workers in the plant.

In view of the fact that the strike had not been officially called off and that a picket line was still at the gates, Mr. Green asked me to permit Mr. Daniel S. Ring, my personal Industrial Relations Officer, to appear before the Executive Committee of Local #16 to explain the new situation which had arisen because of the seizure of the plant by the Navy. Mr. Ring had been borrowed from the Maritime Commission where he was Director of Labor Relations. His presence at the plant was due to the interesting situation down in

Washington where the War Production Board was run jointly by Mr. Knudson and Mr. Sidney Hillman.

During the meeting of Mr. Ring with Mr. John Green and Local #16, Mr. Green informed the members of the Executive Committee that all the issues for which the Local had voted had been promised by the Government. Mr. Ring interjected to state that the instructions with respect to the reopening of the plant were as stated and that he could not, by his silence, allow the assumption that any arrangements for operation were on a broader basis. The statement of Mr. Ring is particularly important because, as we shall see later on, the question of Union Maintenance was continually arising. The union accepted the Navy's statement, the picket line was called off, as well as the strike, and full cooperation was extended to the Navy management.

During the discussions prior to the seizure of the plant, Mr. Hillman had insisted that Mr. Ring should be his representative and that the plant should be jointly operated by the Navy Representative, myself, and Mr. Hillman's representative, Mr. Ring. In a session in Secretary Knox's office at which Mr. Hillman and I were present, Secretary Knox had told Mr. Hillman that the Navy Department would not seize the plant unless it had assurances from Mr. Hillman that his representative, Mr. Ring, would report to me for duty. Mr. Hillman did not quite categorically agree to this demand of the Secretary, but his reply was couched in such terms that the Secretary felt warranted in proceeding. I told the Secretary of the Navy that I had no intention of being jointly in charge of anything with anybody. However, I had no objection, per se, to Mr. Ring as virtually labor's representative. Intercourse between Mr. Ring and Mr. Hillman was always free and untrammelled as far as I was concerned, but this situation could have proved explosive if it had not been for the fine understanding, high character, and competence of Mr. Ring.

This was the Navy's first seizure and operation of a plant. The Army had, a few months previously, seized and operated the plant of the North American Aviation Company, but the operation was of short duration and without the use of Government money. The situation at Kearny was entirely different and was an operation of a fundamental nature.

The Secretary of the Navy left it up to me as to whether or not I should wear a uniform. I decided to wear it. I also selected my own title of "Officer-in-Charge" as being descriptive and not too high flown. This title was used in all succeeding seizures by the Navy.

There were many other things to settle, such as letterheads, handling of Government funds, the use of Government or commercial procedure in purchasing, questions of insurance compensation, cooperation with municipality, county, and state with respect to taxes and many other sorts of dealing which are not encountered in Navy Yard work. In fact, it is true to say that it took over three months at Kearny to make as much progress as we made in three days in the case of later seizures and operations. Hanging over my head all the time was whether or not the Comptroller General of the United States would approve all the unusual expenditures I was making of Government funds.

On 25 August 1941, shortly after 8:10 a.m., I addressed the supervisors and foremen and explained to them that the purposes of the Navy occupation were to restore production. I also explained to this group that the Navy management must protect the interests of the Federal Shipbuilding and Dry Dock Company at all times, subject to the fact that production must go on in the Government's interest.

Considerable trouble was had with the company in regard to their books and records. The former management was finally convinced, however, that under the Executive Order the books and records of the company were under the custody of the Officer-in-Charge but that they were available to company officials for the protection of the company's interests.

On 26 August 1941 the first shift went on duty at 8 a.m. with over 97% attendance, in spite of all the predictions of the company officials that the opening would be a flop.

An agreement was reached between the Officer-in-Charge

and the former management in regard to what officials of the company would remain company officials and what officials of the company would work under the direction of the Officer-in-Charge.

One of the great difficulties in setting up the Navy management was the lack of administrative space for the company, let alone for the company and the Navy. The company had made no attempt, by August 28th, to provide the Navy with an administrative suite, without which an organization or business cannot be successfully run. There appeared to be considerable room on the second floor of the Administration building for the remaining company officials. Therefore, one night I moved the furniture required by the remaining company officers, including the President, to their new location, and the next morning the Navy management occupied the original executive suite and the company officials occupied the space assigned to them on the second floor. At no time were company officials denied access to the Administration building.

Officials from the Reconstruction Finance Corporation visited the plant to appraise the property. It was not known at this time whether or not the Government would avail itself of U.S. Steel's offer to sell the Federal Shipbuilding plant to the Government. Of course, from the Navy angle, it was impossible to tell whether the steel company wanted to get rid of the plant in order to stop the penetration of Union Maintenance among its vast holdings or whether they thought there was a good opportunity to unload a plant which for so many years had been in the "red."

I started out with the idea that the seizure and operation of a plant might easily terminate in a legal dispute in the courts in a suit for damages. For this reason I made every effort to improve the position of the plant as a business in order to lessen the possibility of an adverse decision in any possible court action as far as the Government was concerned. With this end in view, I decided that every effort would be made to increase the earnings of the company.

There were two obviously weak shops in the yard, the rivet

shop and the welding shop. It was reported on September 2nd that there was every evidence of a slow down in the riveting shop, and it was determined to make an example of this shop and to take care of the welding shop later. The matter of increased production in the riveting shop was taken up with the union, and the union assured the Navy management that all limitations on production would cease. As time went on and labor got better acquainted with the Navy management, the amount of riveting rose beyond all previous records and attained heights that had not previously been dreamed of. It should be noted, however, that the increase in riveting was also due to the increase in the number of employees in that particular shop.

The work in raising production per man in the welding shop took a long time but was finally accomplished. The improvement was due to my persistence and my frequent threats to the foreman to take over the shop and run it myself if he couldn't run the shop the way I wanted it run.

Continual adjustments were made in the wage schedules to carry out previous agreements and, while we worked long and hard in trying to cut down the total number of grievances, we finally came to the conclusion that there would always be grievances and that the list could never be cleared up by any particular time.

I was very fortunate in being assigned to the Federal Ship Yard for my first seizure. I was familiar with the layout of the Yard and had an acquaintance with many of the officials of the Yard, from the President, Mr. Lynn Korndorff, down. They were, as a rule, of a very high caliber. This Yard over the years had kept up and improved the so-called "assembly system." Mr. G. G. Holbrook, the General Yard Superintendent, had developed this method while at Quincy, Massachusetts, with Bethlehem, and he introduced it at Federal after World War I and perfected it still further. Much of the work of actual construction was done in erection shops, and it was a common sight to see sections of a ship, weighing 40 or 50 tons, being hoisted up and bolted in place. This was prior to similar excellent work along these lines by Henry J. Kaiser.

The Yard had long employed the berth system. In this system, a ship, for purposes of new construction, is divided into, let us say, "chunks," each "chunk" constituting a natural division of the ship. The work in each "chunk" was planned to be performed by the various trades in sequence, with the end that each trade could work without interference from the preceding trade or the following trade. In other words, when a 40-ton composite was hoisted in place on a vessel building, this section was, in so far as possible, completely equipped with piping, electrical wiring and whatever apparatus belonged to that particular "chunk"—even to the painting.

With a Yard so well organized, and with such competent shipbuilding officials, there was no difficulty for proper leadership to begin to show surprising results in a short time.

The light cruiser, *Atlanta*, was launched on 6 September 1941, and additional launchings proceeded at a planned rate which was constantly being accelerated.

In order that there would be no argument or disagreement at the time when the plant would be finally turned back to its owners, conversations were early initiated with the auditing firm of Arthur Andersen and Company, Certified Public Accountants, to make an audit of the books and records of the company in order to determine, through an independent third party, the correct book values of the accounts involved in the Navy's taking over of the plant. As a result of long conversations, an agreement on the part of the Secretary of the Navy and the company was reached whereby a joint audit was conducted by Arthur Andersen and Company, representing the Navy, and Price, Waterhouse and Company, representing the owners.

By 12 September 1941 a review of the riveting production was made; and it was found that, for the week ending September 10, the drive of rivets was 120,636—the second highest for the year and approximately 30,000 in addition to the average weekly production for the period from June 4 to August 6. This compared with a rivet drive of 61,507 for the week ending September 3 and 63,851 for the week ending August 6.

By 13 September 1941, there was evidence that there was an agitation for an independent union, probably encouraged by some company officials. This agitation was accompanied by the unauthorized circulation of petitions. Naturally such an overt act, contrary to previous working rules, resulted in the reissue of working instructions in a new order known as General Order Number 2.

Every ship yard, or every other manufacturing plant for that matter, has certain rules and regulations in regard to the conduct of employees based on the interests of the owner, the interests of safety, and the best interests of all concerned. This General Order Number 2 merely represented such a customary order and merely reinforced previous instructions. It became obvious that the original instructions had never been enforced and naturally General Order Number 2, which set out to enforce them, must experience some adverse reception; but, on the whole, things gradually quieted down and production continued to mount.

The Navy was treading a precarious path among the various groups which did not always approve of the Navy's actions, since naturally each group was looking out for its own interests. The company officials represented one point of view, labor another, the independent union a third, and the Navy a fourth. In the background was the threat of the Mediation Board injecting itself into the picture by enforcing Union Maintenance.

The press of New York City was very much interested in the operation of the plant. As the Navy management showed improved operating results, the left-wing newspapers were gleeful because they regarded it as additional proof that Government management was superior to private management, and the right-wing newspapers were constantly seeking to prove that private property and private owners were being damaged. Fortunately, Mr. Frank Mason, Special

Assistant to the Secretary of the Navy, had already warned me of this situation. Also, at Mr. Mason's suggestion, I obtained a direct wire to Washington, and this was a great help.

Conferences were frequently held with representatives from the New York press and magazine publishers, ranging in political affiliations from the *Wall Street Journal* and *Business Week* to *PM*. The conduct of these press conferences, in view of these circumstances, became a source of enjoyment.

One of the early difficulties in the Yard which engaged my attention was the high death rate. It could be proved statistically that the longer a new employee stayed in the Yard, the longer he would live. An intensified drive was put on to better the instructions given to new employees with regard to the hazards of a ship yard. Many of the new employees had come from what we might call "protected occupations" and had no previous experience in working out of doors and under conditions which can be hazardous if all the safety rules are not observed and if a man is not wide-awake at all times.

The services of the safety engineer of the New York Navy Yard were early obtained in order to stimulate, inspect, and observe better safety rules at the Federal Yard.

Representatives from the New York Navy Yard inspected the medical and dispensary facilities of the Federal Yard and stated that they met with their approval.

In order to stimulate interest in the riveting and welding shop in their production, the foreman of the welding shop and the foreman of the riveting shop were required each week to come up to my office and plot the new spot on the production curve of his shop. This was a very effective move in stimulating production. Formerly production reports were published monthly and were dead and useless by the time they were made. Keeping graphical track on what the shops were accomplishing had a beneficial effect on them, and any falling off of production was noticeable before it was too late to do something about it.

I decided that I did not care for Navy publicity, with

headquarters in New York, so my publicity was handled by the late Frank Sinnott, distinguished Editor of the *Newark News*.

On 21 October 1941 the Sinclair tanker, Hull #191, was launched. This date had been scheduled before the strike, so that it appeared that sufficient progress had been made during Navy management to offset the effect of the strike.

The method by which we put into effect the purchasing of material by using commercial methods instead of the Navy methods is illustrated by the following interchange of correspondence between me and my Disbursing Officer.

On 25 August 1941, in a letter to my Disbursing Officer I stated:

"I hereby direct and order you to make such payments as may be necessary upon vouchers certified by the Navy Comptroller of the Federal Shipbuilding and Dry Dock Company.

"The provisions of the Executive Order authorizing that all things necessary or incidental to the operation of the plant under emergency conditions precludes the observance of normal Navy disbursing methods and of necessity you will follow the existing Federal Shipbuilding and Dry Dock Company's accounting and disbursing procedures except that disbursements made by you will be in cash or by check through your official Navy account which has been established in the Treasury of the United States for that purpose.

"All disbursements made by you pursuant to the foregoing shall be covered by public vouchers . . . and you shall render an Account Current with such supporting vouchers as may be made available to you by the Navy Comptroller. It is directed that a certified copy of each public voucher paid by you be retained with the files at the plant for use in any future litigation which may arise incident to the government operation of this plant."

The Disbursing Officer wrote back to me on 25 August 1941:

"Despite the conditions outlined in your orders under which the Federal Shipbuilding and Dry Dock Company is to be operated pursuant to Executive Order dated 23 August 1941, I feel constrained to protest the payment of all vouchers presented to me for that purpose by the Navy Comptroller for the reason that such vouchers will not be supported by the usual substantiating papers required under normal Naval disbursing procedure, and those required by law.

"I therefore respectfully protest, under the authority of article 1748, U.S. Navy Regulations, the procedure you have directed me to follow in the conduct of disbursing matters in the operation of the Federal Shipbuilding and Dry Dock Company."

I replied:

"I have considered the implication of the order, directing a disbursement procedure to be followed in the operation of the plant and facilities of the Federal Shipbuilding and Dry Dock Company and since the procedure is imperative to the resumption of normal operations as directed by the President of the United States the order is herewith reiterated and you are directed to carry (it) into operation forthwith."

Working conditions during the period immediately preceding the seizure are best described from a report submitted by the General Manager in which it is stated that labor relations and disciplinary control had reached the point of an impossible status and, in the opinion of the General Manager, "Control of discipline impossible by supervisors until matters definitely settled."

One of the steps taken by the Navy management was to issue Order Number 2, previously referred to, informing employees of the working conditions and then vesting "disciplinary control" through "self government" in the hands of both the workers and the supervisors. In applying the rules under the Navy management, impartial hearings were encouraged to avoid harsh and what appeared to be faulty interpretations of working conditions. Supervisors were cautioned to avoid "self-made rules" or untenable interpretations of the rules contained in Order Number 2.

Mr. Peter J. Flynn, President of Local #16 of the Industrial Union of Marine and Shipbuilding Workers of

America, in a letter dated 31 October 1941 to me, stated that:

"Where any such changes (changes in shift schedules) are contemplated we suggest that the representative of the Union be called into consultation, and that any intimation or general notice in the yard, carry the designation of the Union as an indication that such changes have been arrived at in a manner consistent with the recognized principles of collective bargaining."

In reply to Mr. Flynn's letter, I stated:

"My instructions from the Secretary of the Navy in regard to the operation of the Kearny plant do not authorize me to recognize any Union or group of employees working in the plant as a collective bargaining agency for the employees."

While I did not recognize collective bargaining, I did dispense industrial justice.

While I did not recognize the union, my relations with the union leaders were always cordial. Peter Flynn was always solicitous about my welfare, particularly because of my habit of circulating around the yard when thousands were changing shifts. As I used to say afterwards, Korndorff recognized the union but didn't meet it socially, while I met the union socially but did not recognize it.

In my work in Navy Yards, I made a regular practice of visiting the foremen in their shop offices, whether I had any business to discuss or not. I continued this practice at Kearny. The foremen were delighted and told me that to get visited by the "boss" was a new experience in their lives.

Reclassification of wages was made during the occupation, but no general wage increases were made.

A new system was installed for keeping track of the steel yard.

During the fall, because of union complaints, Mr. Will Davis, Chairman of the National Defense Mediation Board, directed the Board to conduct an investigation of the Federal Shipyard to determine whether or not there was lack of compliance with Union Maintenance. As a result of this investigation, Mr. Davis sent me a copy of the investigation and an order to discharge seven—I think it was—employees for lack of compliance. I took this report to Secretary Knox, saying that only he or the President had authority to order me to discharge these men and that I could not take orders from the Mediation Board. I further suggested that he put the papers in his desk and forget them, which he did.

Other than this, no attempt was ever made to put Union Maintenance into effect at this yard during the Navy operations.

How well did the management record of the Navy compare with the management record of the company? The following schedule shows what the annual production would have been, based on the rate of production of the month indicated:

Navy Operations		*Company Operations*		*Increase per cent*
Oct.	$102,382,865	May	$78,728,410	28.41
Nov.	105,677,355	June	86,398,420	22.31
Dec.	122,506,045	July	88,186,555	38.92

A further comparison leaves no doubt concerning the efficiency of the Navy management:

	TOTAL		NAVY OPERATIONS		COMPANY OPERATI	
	Amount or Number	*Per Cent*	*Amount or Number*	*Per Cent*	*Amount or Number*	*P Ce*
Elapsed Time	365 days	100	129 days	35.34	236 days	64
Net Sales: Ship Contracts Completed	$69,423,716	100	$41,613,561	59.94	$27,810,154	40
Net Operating Profit	$ 6,738,781	100	$ 3,972,570	58.95	$ 2,766,211	41

This improvement of course depended partly on increased employment.

Shortly after Pearl Harbor, I informed Secretary Knox that the time was ripe to turn Federal Ship back to the owners. Everything there was going fine; we had increased production and profits; we had kept out of the Union Maintenance controversy, and it seemed best to get out while the going was good.

The Secretary agreed, and on 6 January 1942 the Yard was turned back to the owners. In the settlement with the company (U.S. Steel), the government retained an amount equal to cost of operation, including officer salaries, etc., but otherwise returned all accumulated profits to the owners.

Much of the success at Kearny was due to the fact that I was fortunate in having unusually capable assistants. Mr. Ring, labor adviser, I have already mentioned. Norwood P. Cassidy, Navy Comptroller, introduced changes in fiscal control, many of which were retained by the company after the plant was turned back. Frank W. Marshall, Special Assistant in the Bureau of Supplies and Accounts, watched the accounts. Commander George A. Holderness (later Rear Admiral), Naval Inspector at this shipyard, was also most helpful always. Captain Orville D. Foutch, SC, USN, was an excellent Disbursing Officer.

Among those who were former company officials were W. C. Hemingway, General Manager. He was in a tough spot, and he performed with distinction. G. G. Holbrook, General Yard Superintendent, and A. R. McMullen, Engineering Superintendent, should also be mentioned most favorably.

Seizure of the Bayonne Plant of the General Cable Corporation

On Thursday, 13 August 1942, the Under Secretary of the Navy, Mr. Forrestal, called me to his office at 6:00 p.m. He told me that I was to seize the Bayonne Plant of the General Cable Corporation at Bayonne, New Jersey, on the following morning, that I was to use troops if necessary to break the wild-cat strike, and that before I left Washington I would report my mission to General Somervell of the Army.

With these oral instructions, I started getting my staff together. I was fortunate in acquiring the services of Mr. Norwood P. Cassidy, Special Assistant to the Paymaster General of the Navy, Mr. Daniel S. Ring, Labor Counselor of the Maritime Commission, and Mr. John Vincent, Special Assistant to the Under Secretary of the Navy, all of whom had been with me in the seizure of the Federal Shipbuilding Corporation at Kearny, New Jersey.

It was imperative to leave that night by train. I wished to do so because we had had notices and orders printed which would be immediately posted on gates, entrances, and walls upon our arrival at the plant.

During wartime, procuring transportation on such short notice was quite a problem indeed and in this connection I can never forget the effect our orders had on a Marine sergeant who was in charge of Mr. Forrestal's reception room. The first response of the Pennsylvania Railroad was that no transportation was available in spite of the fact that we were carrying out the direct orders of the President of the United States. The railroad's decision appeared to be final but when the Marine sergeant began talking over the telephone about the Government's taking over the Pennsylvania Railroad and all other railroads because we couldn't get satisfactory service out of them in an emergency, the railroad thought possibly he had a point, and at 2:10 a.m., we left for Bayonne, New Jersey, on the Pennsylvania.

Before I left Washington I complied with the oral order of Mr. Forrestal and reported to General Somervell that I was about to leave Washington to break the strike at the plant of the General Cable Corporation. General Somervell asked me if I wished a bodyguard. I told him that I did not, and he seemed quite surprised. He informed me that an officer from Governor's Island would report to me for duty and that troops would be available for my use in carrying out the Executive Order. He added that troops at Governor's Island were under alert; that a contingent of the U.S. Army was encamped on the ball park in Jersey City; and that troops had been detrained at Elizabeth, New Jersey, and would be held in readiness to support my efforts at the plant of the General Cable Corporation.

We arrived at the plant of the General Cable Corporation at 8:55 a.m., Friday, 14 August 1942. Brigadier General R. K. Robertson of the Army from Governor's Island reported to me for duty. There was no sign of any disturbance at the plant nor had there been any. The gates were closed and there was a large gathering of people across the street from the main office.

Promptly upon my arrival Mr. Dwight R. G. Palmer, President of the General Cable Corporation, members of the corporation's executive staff including the plant manager, myself and staff, and Brigadier General Robertson met in conference.

Mr. Palmer was most cooperative. He issued orders to his executives that my orders were final and conclusive in the plant during the Government's occupation and operation of the plant and that any expenditures I might order to be made were satisfactory to him. This meant, of course, that the plant would continue to be operated by company, not Government money.

At the end of the conference the General struck an unfavorable note by telling me that he wanted it understood that if the Army took any action against a mob pursuant to my order that he would be in charge of the operation. In deference to good manners which the good Brigadier General apparently lacked, I did not reply to that question in public for the time being.

After the conference broke up, I sent for the General and told him that I would never issue an order to have troops fire upon American citizens unless it would be in defense of life or property or in order to carry out the Executive Order, that I agreed that if combat started the Army should be in charge, but that if I ever gave the order to fire upon a mob, I wanted it understood that there would be no firing over the heads of the mob. He would use ball cartridges and not blank cartridges. After I had stated my position, my relations with the General were enthusiastic and cordial.

In accordance with the practice which we developed at Kearny, copies of the Executive Order which directed the seizure and the Notice of Possession were posted throughout the plant as soon as possible after our arrival.

I was very much surprised later when I saw soldiers around the vicinity of the plant. I had been given to understand that the only troops were at the three locations noted above. I immediately ordered the General to withdraw all troops from the vicinity and to keep them away. My order was promptly complied with. My reason for this order was based on common sense. This was a wild-cat strike, not a strike under labor union supervision. The employees had struck in protest against a decision of the War Labor Board, refusing them an increase in wages. Under such circumstances, with soldiers and employees in too close proximity, there is always the possibility of someone—it doesn't have to be an employee—throwing an empty pop bottle at the soldiers, thereby creating conditions which could rapidly get out of control. Furthermore, I wanted to avoid all evidence of employing force as long as possible.

In further accordance with the procedure we had inaugurated at Kearny there was posted a notice to the effect that "Under Navy management the wage rates, hours of work, overtime, and general working conditions in effect at the time of the cessation of operations will prevail."

In all my seizures of over 100 plants during the war under eight executive orders, we never allowed anybody to profit by striking or cessation of work.

Shortly after the conference with the General Cable Corporation officials, it seemed to me that the best thing to do for a starter would be to open the plant as rapidly as possible. A notice was promptly posted that the plant would be opened at 3:40 p.m. on the same day, Friday, 14 August 1942, to permit returning employees to enter the plant preparatory to resuming work at 4:00 p.m. This information was also broadcast and employees living at distant places were notified by telephone or telegraph.

The employees returned in a most orderly fashion and at 4:20 p.m., 285 men out of the possible attendance of 290 were at work. The plant manager reported at 4:30 p.m. that all operations were proceeding in a normal and very satisfactory manner. The operations on Saturday, 15 August 1942, and thereafter for that matter, continued normal and without incident. On Sunday we were shipping out cable from the plant to be used in the war effort.

It did not take long to realize what had happened. There had been a relatively great increase in the number of employees in this plant. So many new faces were added, that the officials of a good and conservative AFL union had lost contact with their members and, therefore, lost control. This was also true of management. The increase in the number of employees had been so great that management from the president down no longer had the time to get sufficiently acquainted with a large enough number of supervisors to display some of the characteristics of leadership.

At the suggestion of the Office of the Secretary of the Navy, my staff prepared a telegram to be sent by the Secretary of the Navy to Mr. Michael Petrakian, chairman of the employees' Strike Committee. This telegram to Mr. Petrakian was released to the public press and had a most salutary effect, since it convincingly advised the strikers and informed the general public that "Seizure of the Plant was necessary to insure resumption of essential production and must under no circumstances be construed as an endorsement of the attitude of the group of employees who precipitated such action." The newspapers in the New York area commented very favorably on the firmness of the Government's position as conveyed in the telegram.

The positive declaration that the Government's possession was not to be considered as condoning the action of the strikers, and the further fact that upon arrival at the plant, I had posted a notice to the effect that no benefits, which did not exist prior to the strike would accrue to the employees under Navy management, immediately produced a strong realization among all employees in the plant that nothing was to be gained through the continued possession and operation of the plant by the Government.

In view of these facts, I directed Mr. Daniel S. Ring, my personal industrial relations adviser, to get in touch with Mr. James J. McNally, President, and the Executive Board of the local A.F.L. Union, I.B.E.W., Local #B-868, and to request them to call a meeting of the employees of the plant so that they could consider a resolution disavowing the strike, agreeing to recognize the decision of the War Labor Board, and further abiding by the principles of collective bargaining, thereby insuring to the Government that there would be no further interruption to production throughout the duration of the war. The president of the Union and the Executive Board approved the resolution we suggested to them and they agreed to call a Union meeting, but suggested it be held on Friday evening, 21 August 1942.

Realizing the necessity for speedy action, I insisted that the meeting be held not later than Tuesday evening, August 18th. I overrode their objections which were based on the fact that their members lived in widely separated areas and were difficult to reach promptly. They finally consented to the earlier meeting of the union.

On the morning of Tuesday, August 18th, following the receipt of the telegram from the Secretary of the Navy, Mr. Petrakian requested an appointment with me, which I refused in order to deny official recognition of the strike committee. However, it was arranged for him to confer with Mr. King and Mr. Cassidy of my staff.

At this conference, it developed that Mr. Petrakian felt that his position had become untenable. He was a much worried man. He stated that he had no intention of defying the Government and was anxious to do whatever he could to assure the Navy of this point. I recognized that the wildcat strike was an emotional outburst and that no one was more surprised than Petrakian when events which he admittedly set in motion propelled him into the unsought position of leader of the wild-cat strike.

He was advised, first, to send a telegram to the Secretary of the Navy informing the Secretary of his present attitude; second, to support the resolution to be offered by the Executive Board of the union in the meeting to be held that evening. To all of this he agreed. My staff assisted him in preparing his telegram to the Secretary of the Navy.

Meanwhile, my lawyer, the company lawyer, and Mr.

Dwight R. G. Palmer, had been in conference in New York City in the main offices of General Cable. After their conference was completed, they joined me in Bayonne. Imagine their amazement when I told them that the strike was all over—and it was.

On Tuesday evening, August 18th, only a few hours later, the union meeting was held in the Polish Hall in Bayonne, New Jersey, with an attendance of approximately 600 of the 817 employees of the plant. The meeting was conducted under the auspices of the union and was addressed by Mr. Alexander Smalley of the Labor Division of the War Production Board, Mr. William Walker, District Vice President of the I.B.E.W., and Mr. Daniel S. Ring. Mr. Norwood P. Cassidy, my fiscal consultant, was also present at the meeting.

After the addresses were concluded, and upon the motion of Mr. Michael Petrakian himself, the resolution of the Executive Board was unanimously adopted. This action of the employees effectively removed the necessity for further continuance of Navy possession and operation of the Bayonne Plant of the General Cable Corporation, and this fact was reported to the Navy Department.

On 20 August 1942 the President of the United States signed an Executive Order terminating the Navy's possession and operation of the plant. General Cable did not claim any reimbursement from the Government due to any expense to the company because of Navy seizure and operation.

My Assignment to the Office of the Under Secretary of the Navy

On 5 November 1942, having been detached from the Naval Research Laboratory, I reported to the Under Secretary of the Navy, Mr. James V. Forrestal, as Special Assistant. Some months previous to my detachment from the Laboratory, I had been relieved of my additional duties as Technical Aide to the Secretary of the Navy and Head of the Office of Inventions and replaced as the Naval representative of the National Defense Research Committee.

Secretary Knox, at this time, gave me the only unsatisfactory fitness report I have had in my long Naval career. He charged that I could not "get along with civilians."[1] Admiral Nimitz, Chief of Naval Personnel, did not agree with the Secretary and attempted to persuade him to change the report, but to no avail. Under the circumstances, and also because Forrestal was so likeable, I was delighted with my new assignment.

During my plant seizure of Federal Ship and the strikebreaking at Bayonne, I received several commendatory letters signed by Secretary Knox. These laid especial emphasis on my ability to "get along with civilians." I thought perhaps Knox was suffering from a guilty conscience, but years afterwards I discovered that the letters were framed by the Secretary's personal lawyer, a Special Assistant to the Secretary, a man for whom I have always had the highest regard. The gentleman who helped me anonymously when I needed a friend was none other than the recent Presidential candidate of the Democratic party, Adlai E. Stevenson.

Captain (now Vice Admiral, Ret.) Morton L. Deyo and Captain (now Vice Admiral, Ret.) Frank E. Beatty tried, as Aides to Secretary Knox, to bolster my fortunes. In this they were enthusiastically, if unsuccessfully, assisted by Captain (now Rear Admiral, Ret.) James E. Maher. Someone recommended me to the Awards Board for a medal, but the award was whittled down to a Letter of Commendation, to wit:

LETTER OF COMMENDATION—SECRETARY OF THE NAVY

For outstanding services in the development and perfection of Radar while Chief of the Bureau of Engineering and later as the Director of the Naval Research Laboratory. Through his tireless efforts and devotion to his cause, Rear Admiral Bowen convincingly presented this new weapon of war to the commercial concerns qualified to

[1] Actually the Secretary said: "Very capable engineer and possessed of great courage in pressing experimental projects. Rendering exceptional service in research work. Rather belligerent and temperamental in his contacts outside Navy."

produce it and, by sponsoring a wider understanding of the merits of Radar, contributed in great measure to the expansion of our facilities for manufacturing this type of equipment in the large volume necessitated by the outbreak of war.

I had not been in Mr. Forrestal's office very long when an order, dated 3 December 1942, issued by the Bureau of Naval Personnel, required that all officers 56 years old or older should be examined physically during February 1943. As a result of this examination, I was ordered before a physical examining board. Was I hopping mad! I started my investigation with the Surgeon General and was referred to a Medical Officer at the Bethesda Naval Hospital for information. I found out that I was charged with everything under the sun, beginning with "alopecia" (baldness). Finally, the medical officer admitted to me that all the alleged defects only added up to the fact that I was 59 years of age. My feelings were further exasperated by the rumor that the Navy Department wished to retire two Rear Admirals, and that it used this method to get rid of them quietly. The rest of us were just expendable in the purge.

Secretary Knox was out of town the day I decided to have it out with the Navy Department, come what might.

I went into the Under Secretary's office and did some of the plainest speaking I ever did to any superior in my 46 years in the Navy. I called attention to my successful handling of the delicate job at Federal Ship and to the fact that I had just finished breaking the wild-cat strike at Bayonne, without once using troops and disdaining a bodyguard. I demanded to know if such performance didn't outweigh the results of a physical examination. I said that if the Department were insistent on a purge, I could nominate some real stuffed shirts. As a parting shot, I told him I had no intention of becoming retired and that I would fight my way down Pennsylvania Avenue, through the White House, to Congress.

Forrestal was very sympathetic. He told me that this was the first he had heard of the purge.

The retiring board before which I appeared in February 1943 was headed by Rear Admiral Dallas G. Sutton, MC, USN. The board found me physically fit to perform active duty, even with "alopecia."

In 1947, having become due for retirement, I tried every means possible to get a retirement based on physical defects, instead of age, but with no success. A physical retirement would have exempted my retired pay from income tax. But no, it seems that I was terribly healthy.

Seizure of the Plant of the Howarth Pivoted Bearings Company, at Holmesburg, Pennsylvania

EXECUTIVE ORDER

AUTHORIZING THE SECRETARY OF THE NAVY TO TAKE POSSESSION OF AND OPERATE THE PLANT OF THE HOWARTH PIVOTED BEARINGS COMPANY AT PHILADELPHIA, PENNSYLVANIA.

WHEREAS Howarth Pivoted Bearings Company has entered into contracts for the manufacture of war materials essential to the construction programs of the United States Navy Department and the United States Maritime Commission, and

WHEREAS it is deemed essential that the plant of Howarth Pivoted Bearings Company, located at Philadelphia. Pennsylvania, be taken over for use and operation by the United States of America in order that it may be effectively operated in the manufacture of the kind, quantity, and quality of such war materials;

NOW, THEREFORE, I, FRANKLIN D. ROOSEVELT, pursuant to the powers vested in me by the Constitution and laws of the United States, as President of the United States and Commander-in-Chief of the Army and Navy of the United States, hereby authorize and direct the Secretary of the Navy immediately to take possession of and operate the plant of Howarth Pivoted Bearings Company, located at Philadelphia, Pennsylvania, in order to produce effectively essential war materials required by the United States of

America and to do all things necessary or incidental to that end.

The Secretary of the Navy may exercise the authority herein conferred through and with the aid of such person or persons or instrumentalities as he may designate, and may select and hire such employees, including a competent civilian advisor on industrial relations, as are necessary to carry out the provisions of this order, and in the furtherance of the purposes of this order, the Secretary of the Navy may exercise any existing contractual or other rights of the said company, and take such other steps as may be necessary or desirable.

Possession and operation of the said plant under this order will be terminated by the President as soon as he determines that such plant will be operated privately in a manner consistent with the war effort.

/s/ FRANKLIN D. ROOSEVELT

THE WHITE HOUSE,
June 14, 1943

14 June 1943

Dear Admiral Bowen:

By Executive Order dated 14 June 1943, a copy of which is enclosed, the President of the United States directed the Secretary of the Navy immediately to take possession of and operate the plant of Howarth Pivoted Bearings Co. located at Philadelphia, Pennsylvania, and, insofar as may be necessary or desirable to produce the essential war materials which must be obtained from such plant, or as may be otherwise required for the war effort and to do all things necessary or incidental to that end.

Pursuant to the foregoing direction, I, as Secretary of the Navy, hereby designate you to act for me in carrying out the duties and responsibilities imposed upon me by said Executive Order.

Accordingly, you will take immediate possession of the plant of Howarth Pivoted Bearings Co., including all of the real estate, buildings, facilities, machinery, tools and equipment, and all materials, supplies, and articles of production whatsoever at the plant or elsewhere essential for the efficient operation of such plant. You shall also assume possession of all pertinent records, books of account and other documents pertaining to the company's operation of its plant, and, to the extent you in your absolute and uncontrolled discretion may deem proper, you may make suitable arrangements with such official of such company as may be approved by you, whereby he may, subject to your supervision and control, have access at reasonable time to such records, books of account and other documents.

You will inform the local representative of the company and the employees of the contents of the Executive Order and of these instructions. Copies of the Executive Order and notice that you have, pursuant thereto, taken possession of the plant should be posted in conspicuous places on the plant premises.

You will cause operations at the plant to be continued without interruption or delay. The war materials called for by the contracts of the company must be produced without delay in accordance with the terms of such contracts, subject to any modification or changes therein in which you may concur.

In such manner as may be practicable without interrupting operations, you will take full and complete inventory of the plant and of all work done or begun in, upon, or about the materials and equipment there in course of production and assembly and of all materials on hand or on order applicable thereto.

You will employ without regard to civil service or classification laws or regulations all persons and agencies necessary to carry out these instructions, including to the extent you desire the executive officers and employees of the company. You are also authorized to employ the services of any association, firm, company or corporation in furtherance of said Executive Order and these instructions. You will observe the wage rates and hours of employment of the employees of the plant which were prevailing immediately prior to your taking possession of the plant and you may, as permitted by the Executive Order, employ a competent civilian adviser on industrial relations. Fiscal officers will be assigned at once.

You are authorized to procure, in such manner as the circumstances warrant, all materials and supplies necessary to carry out these instructions including the assumption of any existing contracts or orders of the company. You are authorized in your discretion promptly to pay any and all subcontractors of, and suppliers of essential materials to, the company for materials delivered or to be delivered to the plant. You are also authorized to procure, by purchase or rental, such machinery, equipment, and tools as may be necessary. If you desire to have the company continue the procurement of materials, equipment, machinery, tools and supplies, you may make suitable arrangements with the company for such procurement, and, in turn, procure either by purchase or rental any of such items.

In taking possession of the plant and its operation, you do not thereby, and need not, assume any existing obligations or commitments of the company, except such as may hereafter be specifically assumed by you. In this connection you will refer all inquiries with respect to outstanding contracts, accounts and obligations of the company, not assumed by you, for labor, materials, supplies, services, or taxes, to the officials of the company for disposition by them. Claims against the company arising out of transactions prior to the date hereof, unless assumed by you, should also be referred to the company.

You will maintain a system or systems of accounting which will record all transactions embraced in the operation of said plant.

Insurers against loss or damage to the properties of the company's plant should be promptly notified by you of the change in the status of the plant.

You will advise me by telephone or telegram, with confirmation in writing, concerning all important matters connected with the operations herein directed. Further instructions will be issued to you from time to time.

Very truly yours,

FORRESTAL

Rear Admiral H. G. Bowen, USN
Navy Department
Washington, D.C.

Operating conditions at the Howarth Pivoted Bearings Company had become so aggravated and production had fallen so low, that emergency measures had become absolutely necessary to prevent a complete collapse of the entire organization. Various measures had been tried on previous occasions, such as advancing funds to meet operating expenses; employing additional managerial talent; urging Mr.

H. A. S. Howarth, president of the company, to arrange for a strengthening of the corporate interests—all of which were intended to aid in finally establishing a sound concern which could accomplish the original purpose of production of satisfactory bearings in adequate quantities. These efforts had not been successful however. This plant was one of two which supplied all the thrust bearings for new ships which were built for the Navy and the Maritime Commission. There is no substitute for such equipment.

Pursuant to Executive Order No. 9351 dated 14 June 1943, and to instructions from the Secretary of the Navy, I seized the plant of the Howarth Pivoted Bearings Company at Holmesburg, Pennsylvania, on the morning of 15 June 1943.

Assisting in the seizure were a Westinghouse group, Lieutenant Murray Hotchkiss, Disbursing Officer, Lieutenant Ethan A. Hitchcock, General Counsel's Office, and Mr. Frank W. Marshall, Executive Accountant and Auditor. Commander R. E. W. Harrison, Assistant Officer-in-Charge, took his party, consisting of Lieutenant Vennerman and Mr. Dalton, an accountant from the Pittsburgh Office of the Westinghouse Company, to the downtown Philadelphia office of the Howarth Company in the Girard Trust Building.

By letter of intent dated 16 June 1943, I engaged the Westinghouse Electric and Manufacturing Company as managing agent, to operate the plant. Mr. Latham E. Osborne, Vice President of the Westinghouse Electric and Manufacturing Company and Manager of the South Philadelphia plant of the Westinghouse Electric and Manufacturing Company, as Managing Agent, appointed Mr. H. A. S. Howarth Chief Engineer and General Sales Manager. By this arrangement Mr. Howarth was retained on the rolls as an employee of the Managing Agent.

When the plant was inspected by me and my party, accompanied by the Westinghouse Management Group, it was observed that much of the machine tool equipment was thirty years old or older; that much of the plant equipment was inadequate and of poor quality; that improper shop methods and procedures obtained, and that work in progress and finished work were not inspected properly.

Steps were immediately initiated for the installation of improved machine tools of standard design, size, power, and capacity considered to be adequate for the work to be undertaken.

Occupancy of the plant by the Navy promptly revealed that the following condition existed:

(a) Shipments were six to eight months behind customers' requirements.
(b) Production load demanded sub-contracting of many parts in order to produce as quickly as possible.
(c) Available machine tools could not produce the close tolerances required.
(d) Continual breaking down of operations to small lot quantities.
(e) Purchase orders for material had been placed with many suppliers in small quantities because of the Howarth firm's very unsatisfactory credit position. It was also forced to purchase orders from small and in some instances inexperienced suppliers to try to expedite shipments. During the early days of Navy management large and reliable suppliers refused to give any consideration to or supply material for our requirements.
(f) Considerable reworking of finished parts found in the inventory was necessary. These parts were discovered to be defective, and, due to the time element required to obtain new castings, rework was necessary.
(g) Twenty-four line shaft bearings for Navy cruisers were fabricated and machined incorrectly, along with approximately 400 thrust bearing shoes.
(h) Fifteen complete aircraft thrust bearings, with their spare thrust shoes, Falk thrust meters and adaptors for cruisers, and a large number of thrust bearings,

shoes and collars were rejected and returned by customers.

(i) Necessary tooling, such as jigs and fixtures was lacking. Also, small tools, such as drills, taps, counterbores, and particularly gauges and measuring equipment were not available for required production.

(j) An expansion of personnel was necessary. The influx of approximately 200 people, 60 of whom were female, made it necessary to train many inexperienced people.

(k) No standard costs were available.

(l) Excessive overtime was necessary to meet schedules.

(m) The efficiency of operators was poor.

I informed Mr. Howarth and his attorney of the changed status under which the plant was now to operate and the manner in which Mr. Howarth would continue as an employee of the Navy-Howarth Operation.

On 16 June 1943, arrangements were made by me with Charles S. Rockey & Company, Certified Public Accountants, to supervise the taking of an inventory and to make an audit of the books and records of the Howarth Pivoted Bearings Company as of the close of business 15 June 1943.

The art of plant seizure, occupancy and operation had so far proceeded by this time that crystallization of technique had already begun to be apparent.

My first "Notice" stated as follows:

"By Executive Order dated June 14, 1943, the President of the United States has directed that the Secretary of the Navy take immediate possession of the Plant of Howarth Pivoted Bearings Company, Holmesburg, Philadelphia, Pennsylvania, through such person as may be designated.

"Pursuant to the Executive Order of the President, the undersigned has been designated to take possession of this plant, and such possession has this day been taken.

"Any interference with the operation of this plant is an offense against the United States."

It took some time before I could get away with that last paragraph. One lawyer said that he didn't know whether the statement was correct or not. Since no Executive Order to seize a plant had yet been tested in court, I felt that I knew as much as any lawyer about what my authority was under an Executive Order. Anyhow it was common sense in my opinion and it worked.

The order in regard to wage rates etc. had been boiled down to: "Under Navy Management, the wage rates, hours of work, overtime and general working condition in effect at the time of possession of this plant will prevail."

The Managing Agent was installed by my order as follows: "By virtue of the authority vested in me by the Secretary of the Navy, the Westinghouse Electric and Manufacturing Company is authorized to manage the Plant of the Howarth Pivoted Bearings Company, subject to my direction."

An exchange of telegrams between John Green, President of the Industrial Union of Marine and Shipbuilding Workers of America and the Navy Department took place shortly after the plant seizure. They are quoted herewith:

PHILADELPHIA PENN 404 P JUN 18 1943

MR. RALPH BARD
UNDER SECRETARY OF THE NAVY WASHN DC

WE QUESTION THE RIGHT OF REAR ADMIRAL BOWEN TO INTERFERE WITH COLLECTIVE BARGAINING BETWEEN LOCAL 35 OF THE INDUSTRIAL UNION OF MARINE AND SHIPBUILDING WORKERS OF AMERICA CIO AND THE HOWARTH PIVOTED BEARING COMPANY OF PHILADELPHIA WHEN OUR NEGOTIATION COMMITTEE WENT IN TO SEE THE MANAGEMENT LAST NIGHT THEY WERE ADVISED BY MR. BECK THE NEW PLANT MANAGER APPOINTED BY THE NAVY, THAT HE HAS RECEIVED INSTRUCTIONS FROM ADMIRAL BOWEN THERE WILL BE NO CONTRACT NEGOTIATED BY THE NAVY WITH THE UNION THIS UNPRECEDENTED VIOLATION OF COLLECTIVE BARGAINING RIGHTS BY ADMIRAL BOWEN PRESENTS A GRAVE THREAT TO PRODUCTION AT THE PLANT AND HAS GRAVELY IMPAIRED THE MORALE OF THE WORKERS. A VERY SERIOUS SITUATION IS DEVELOPING THERE AND THE MEMBERS OF THE UNION ARE GRAVELY CONCERNED WITH DEVELOPMENTS AFTER NEXT

WEDNESDAY JUNE 23 WHEN THEIR PRESENT AGREEMENT WITH THE COMPANY EXPIRES. I STRONGLY URGE THAT YOU INSTRUCT REAR ADMIRAL BOWEN TO ADVISE HIS CIVILIAN MANAGER OF THE HOWARTH CO TO CONTINUE THE NEGOTIATIONS WITH OUR UNION WHICH ARE NOW NEARING COMPLETION AND THAT IN ANY EVENT THE TERMS OF THE EXISTING AGREEMENT BE EXTENDED UNTIL A NEW CONTRACT IS SIGNED.

JOHN GREEN PRESIDENT INDUSTRIAL UNION OF MARINE AND SHIPBUILDING WORKERS OF AMERICA

TO: MR. JOHN GREEN PRESIDENT
INDUSTRIAL UNION OF MARINE
AND SHIPBUILDING WORKERS OF
AMERICA,
534 COOPER STREET
CAMDEN NEW JERSEY

DATE 19 JUNE 1943

TEXT: IN RESPONSE TO YOUR TELEGRAM THE PLANT OF THE HOWARTH PIVOTED BEARINGS COMPANY WAS SEIZED BY THE SECRETARY OF THE NAVY IN ACCORDANCE WITH AN EXECUTIVE ORDER X THE SECRETARY OF THE NAVY HAS DESIGNATED REAR ADMIRAL H. G. BOWEN USN AS HIS REPRESENTATIVE IN THE CAPACITY OF OFFICER IN CHARGE OF THE PLANT X THE WESTINGHOUSE ELECTRIC AND MANUFACTURING CO HAS AGREED TO FURNISH THE NECESSARY MANAGERIAL ASSISTANCE TO REAR ADMIRAL BOWEN X HOURS OF LABOR X WAGE RATES AND WORKING CONDITIONS WERE FROZEN AS OF THE DATE OF SEIZURE COMMA NAMELY JUNE 15, 1943 X THESE CONDITIONS WILL CONTINUE AS LONG AS THIS PLANT IS OPERATED BY THE NAVY DEPARTMENT EXCEPT THAT IF YOU HAVE ANY COMPLAINTS IN REGARD TO WAGES COMMA HOURLY RATES OF PAY OR WORKING CONDITIONS YOU MAY TAKE THEM UP WITH REAR ADMIRAL BOWEN THROUGH HIS AUTHORIZED OPERATING AGENT COMMA MR. BECK X REAR ADMIRAL BOWEN IS AUTHORIZED TO CONSIDER ANY SUGGESTIONS MADE BY THE UNION ON THEIR MERITS X SIGNED RALPH A BARD ASSISTANT SECRETARY OF THE NAVY.

This exchange of telegrams is merely an indication of one phase in a long campaign of education to acquaint labor, management, Bureaus of the Navy Department, and other executive branches of the government with the meaning of an Executive Order of the President of the United States when it directs seizure and operation of a plant in time of war.

On 24 June 1943, arrangements were made by the Supervisory Cost Inspector of the 4th Naval District to appoint a Cost Inspector to the plant of the Howarth Pivoted Bearings Company for the purpose of auditing the claims for reimbursement which would be made by the Westinghouse Electric and Manufacturing Company managing and operating the plant under contract with me.

On 25 June 1943 it was apparent that the Howarth Company was without doubt insolvent, with approximately $344,-000 in assets and $538,000 in liabilities.

On 28 June 1943, a conference was held in my office with Mr. Dechert, attorney for the Howarth Company and Messrs. Hensel, Detmar, Commander Harrison and Lieutenant Hitchcock representing the Navy. Mr. Dechert was offered two suggestions by Mr. Hensel, as follows:

(a) That Mr. Howarth discuss with creditors the possibility of reorganizing the Company in such a manner as to provide good management and make it possible to return the property.

(b) That events be allowed to take their course, which would probably result in bankruptcy.

On 3 July 1943, Lieutenant S. F. Procopio of the Navy Insurance Division, confirmed the conditions with respect to insurance matters with Mr. Hubbel, insurance specialist for the Westinghouse Electric and Manufacturing Company.

On 5 July 1943, Mr. Osborne submitted a list of machine tools deemed necessary by the Managing Agent to operate the plant efficiently. Certain of the machine tools were secured through transfer from other Government activities, while others were purchased. At this time conditions indicated that the procurement of raw materials for continuing operations of the plant would be an aggravating one, due to the unwillingness on the part of suppliers to accept further orders from the Managing Agent because of the unpaid Howarth accounts still on their books. This condition continued until some time in the early part of 1944.

The Managing Agent, on 19 July 1943, at the request of the Officer-in-Charge submitted a report covering conditions prevailing at the date of seizure of the plant. This report not only confirmed the conditions already well known, but further proved the necessity for seizing the property and establishing a sound operating basis under competent management.

On 24 July 1943 the back-log of orders with their quoted prices was analyzed by the Managing Agent, and it was estimated that a loss of approximately $40,000 would be sustained in the production and shipment of bearings at the quoted contract prices that Mr. Howarth had accepted. It later developed that a loss of approximately $125,000 was experienced by producing the bearings on order at the date of seizure.

On 6 September 1943 the Managing Agent was directed to prepare and submit a weekly direct labor summary. This summary marks the beginning of a sound basis for determining operating costs.

On 23 September 1943, Mr. Weisman, attorney in Philadelphia for the Western Foundry Company of Chicago, a vendor of the Howarth Pivoted Bearings Company, was requested by Commander Harrison to arrange for the release of certain patterns which were in the possession of that Company so that production would not be delayed. From this date on, Mr. Weisman became active in representing the creditors of the Howarth Pivoted Bearings Company, and later served as one of the attorneys for the creditors' committee in the Federal Court Receivership.

On 29 September 1943, a Deputy United States Marshal came to the plant for the purpose of serving papers on Howarth Pivoted Bearings Company, a corporation, with regard to a suit entered by Western Foundry Company. He was advised that he should serve the papers on an official of the Howarth Pivoted Bearings Company.

On 4 October 1943, Mr. H. A. S. Howarth came to my office for a conference. Mr. Howarth at this conference requested, as on numerous other occasions, that the Navy Department indicate the conditions which would be acceptable for a return of the plant to him for future operations. Mr. Howarth was again advised that the Navy Department must be assured that a management competent and acceptable to the Navy Department with sufficient working capital to carry on the business be provided before the plant could be returned.

On 5 October 1943 it was reported by the Managing Agent that the standard costs were being established and it was expected that they would go in full operation on 1 November. An inventory of raw materials and work in process was ordered to be taken for the purpose of starting the cost system properly.

Representatives of the creditors of the Howarth Pivoted Bearings Company presented themselves at my office, 29 October 1943. The creditors' representatives were seeking advice as to the proper course of action to take in order to collect their accounts, which had not been paid by the Howarth Company at the date of the seizure of the plant. These representatives were informed that the Navy Department would not be embarrassed by any legal action that they might desire to take.

On 30 October 1943, Mr. Howarth was again advised by me that the Navy Department would lend every effort to aid the corporation in assembling those data needed to substantiate claims for escalation in connection with shipments under contracts that had been executed prior to the occupancy.

On 2 November 1943, at my direction, Lieutenant Murray C. Hotchkiss went to the office (Girard Trust Building, Philadelphia, Pennsylvania) of Mr. H. A. S. Howarth and demanded that blueprints and sales correspondence be turned over for delivery to the plant. Since Mr. Howarth had neglected on numerous occasions to carry out the responsibilities of Chief Engineer and General Sales Manager, the position assigned him by the Managing Agent, it became necessary to demand his resignation. His resignation was accepted.

On 27 November 1943, Mr. Osborne informed the Officer-in-Charge that he was concerned about the inventory value as determined in accordance with the standard costs system recently installed. Current inventory values were unduly inflated because of an entirely inadequate cost accounting system. Until a proper system of cost accounting was completed by which it was possible to allocate indirect and overhead expense to each manufacturing order, it was not possible to determine accurate inventory values. At that time, the inventory values of work in progress were adjusted.

On or about 29 January 1944, receivers were appointed by the Federal Court for the Howarth Pivoted Bearings Company.

By June it became obvious that the operations of the plant had become routine and that I should be free from such responsibilities in order to be able to take charge of future seizures. Accordingly, on 30 June 1944 I was relieved as Officer-in-Charge by Rear Admiral E. L. Cochrane, USN, Chief of the Bureau of Ships.

In order to measure the results during the period from 15 June 1943 to 30 June 1944, with respect to the operations of the Howarth Pivoted Bearings Company under Navy management, the following summaries are presented with brief comment:

(a) Sales or shipments of bearings

By H. P. B. Co.

1943		*1943*	
March	$ 60,000	July	$108,945
April	40,280	Aug.	107,283
May	40,020	Sep.	120,778
June	32,563	Oct.	127,910
Total	$172,865		$464,916
Monthly Average	$ 43,216		$116,229
Resulting Profit or (Loss)			$ 4,588

By Managing Agent

	1943		1944	
	Nov.	$158,555	Mar.	$180,199
	Dec.	162,103	Apr.	176,194
(1944)	Jan.	173,704	May	171,558
	Feb.	174,874	June	160,377
	Total	$669,236		$688,328
	Monthly Average	$167,309		$172,082
	Resulting Profit or (Loss)	($101,391)		$ 68,565

This comparison shows rather conclusively the trend of shipments as well as the profit or loss realized by periods. One of the major conditions confronting the Managing Agent was a lack of adequate accounting control until a complete revision of the accounting system was effected. As of 1 November 1943, the revised cost system was completed and by February 1944 operations resulted in a net profit.

(b) Production and Planning. Since there was no Production or Planning Department prior to the occupancy by the Navy Department, the Managing Agent took steps immediately to organize such a department. Production schedules were set up and shipments went forward under controlled conditions.

(c) Factory Equipment. So much of the equipment was inadequate that in order to produce sufficient bearings according to specifications, I was forced to acquire through transfer when available and by purchase in other instances sufficient machine tools to meet the production schedules.

The most important development in the Howarth seizure was the creation of a new organizational tool—the Managing Agent—who accepted a contract with the Officer-in-Charge to furnish managerial services. The Navy Department did not have available the specialists necessary to reorganize plants, improve shop practice or introduce new ones, set up cost accounting systems, etc. So I talked the Westinghouse Corporation, with the help of George H. Bucher, President, into going along with me in my first experiment with a contract managing agent. I chose Westinghouse because of the proximity of their large establishment at South Philadelphia. Mr. Latham E. Osborne, now Senior Vice President of Westinghouse, then Vice President of the South Philadelphia plant, was amazed to find out that his company could have a Navy Contract to manage the Howarth plant and only have to deal with one Naval officer. That must have helped to deaden the pain. It was a small part of Westinghouse activities, but Westinghouse always realized what an important part of the war effort it was. Mr. Beck, Resident Manager, Mr. D. W. R. Morgan, now Vice President of the South Philadelphia plant, and Mr. Dalton all gave unstinted assistance. At one time even Mr. Roscoe Seybold, Westinghouse Comptroller, paid us a visit and made forthright contributions to our new cost accounting system. No fee was charged by Westinghouse for their management services.

I was most fortunate in having on my staff Commander R. E. W. Harrison, USNR, who was a machine tool designer in civil life. His services were invaluable in shop retooling.

Seizure of the Remington-Rand "N" Division Plant, Elmira, New York

The basic reason for seizing the plant, "N" Division, at Elmira, New York, and terminating the efforts of Remington-Rand, Inc. as a manufacturer of bombsights was, according to the Executive Order, the failure of this activity to produce an acceptable quality of bombsights in sufficient quantity.

The following summary of the production schedule shows that for the six months' period immediately prior to the seizure, actual accomplishment was completely inadequate as compared with potential plant capacity.

Month (1943)	*Production*
June	29
July	30
August	72
September	–
October	–
November	22
TOTAL	153

I had been assured at the time of the seizure by the Chief of the Bureau of Ordnance that the seizure would be of not more than one month's duration. Nevertheless, the Navy operation of this plant under Executive Order dragged on for a much longer period.

It was decided by the Bureau of Ordnance that the proper managing agent for the Remington-Rand plant would be the Carl L. Norden Company, well known producers of the Norden Bombsight. I always prefer to pick my own managing agents, but in view of the fact that I did not have any knowledge of bombsight manufacture, I accepted the nomination of the Bureau of Ordnance and hired the Norden Company as my managing agent.

At 10:15 a.m. on 29 November 1943, I arrived at the "N" Division offices of Remington-Rand, Inc. with my seizure party and immediately proceeded to the office of Mr. A. M. Ross, Operating Vice President, at which time I advised Mr. Ross that the plant had been seized under the authority of Executive Order No. 9399, dated 25 November 1943.

Mr. Ross was very doubtful that I could seize the plant without his communicating with Mr. James Rand, President of the Remington-Rand, Inc. I informed Mr. Ross that I had seized the plant and that it only remained for him to inform Mr. James Rand of this fact. However, I was gratified to learn later that Mr. Rand had announced that there would be no opposition to the seizure and that Remington-

Rand, Inc. would give the Navy Department every cooperation in achieving its objective.

Shortly after the seizure, Mr. Barth, President of C. L. Norden, Inc. and I delivered short talks to the supervisory staff of the plant about the nature and purpose of the seizure. The usual press conferences were held immediately. Later on in the day I, together with Mr. Barth and Mr. Marvelle, the General Manager of the C. L. Norden Company, informed a conference of plant foremen of the installation of the new management and the status of foremen under the changed conditions.

An inspection of the plant, which immediately followed, revealed certain material defects and further that production was almost at a standstill because of the non-delivery of vital materials from one section of the plant to another. In many sections, employees were killing time reading magazines and newspapers. There was nothing else for them to do. Later on in the day I agreed with Mr. Ross with respect to the Remington-Rand officials whom the Navy desired to retain to operate the plant and the officials that were no longer necessary for continued Navy operation. This meeting terminated with renewed expression of cooperation on the part of Remington-Rand, Inc.

On 30 November the various details with respect to disbursements and the validation of vouchers was consummated. This naturally followed from the fact that as a rule, the Navy Department in its seizures operated the plants with Federal money.

Of course, in all these seizures there were many, many questions which arose in regard to retirement fund equity, employment benefits and other matters of similar nature. For the interim period, that is until the future of the plant was decided, it was announced that the Navy would make such appropriate contribution to the Remington-Rand retirement fund as required to protect the employees' interests. Of course there were problems with the cafeteria, which is just one of those things one has to bear in operations of this nature.

One of the things which annoyed me beyond description

and contributed to my lack of confidence in our arrangements was the fact that hardly a month had elapsed before the managing agent submitted to me a schedule of items required to round out and complete the administrative and productive equipment of the plant. The estimate for this work amounted to approximately one million dollars, of which about five hundred thousand was for additional special tools, jigs, gauges, and precision equipment. I was amazed that such a large sum of money should be required to equip a plant which was a brand new facility and which had operated for only six months. I felt as if this condition of affairs should have been discovered and corrected long before the seizure.

It was most unfortunate that I had to seize the plant of the Los Angeles Shipbuilding and Dry Dock Company a few days after I seized the plant of the Remington-Rand "N" Division and, consequently, was not able to give the plant's affairs the personal attention they deserved. However, I left at the Elmira Plant as my Deputy a very competent Naval Reserve Officer, Commander R. E. W. Harrison.

Due to the pressing reorganization problem of Los Angeles Ship, and also due to sickness, I was not able to return to my office in the Navy Department for approximately three months.

While I was in California I had kept in reasonable touch with the officers at the Elmira plant, but none of the information I had received from that source gave me too much confidence that operations were in a satisfactory condition. Consequently, after my return to Washington on 1 March, I arranged what I called a "town meeting" in order to get together comments from all the disagreeing elements involved in the Remington-Rand occupation. Specifically, all interested parties attended this meeting except Remington-Rand. Perhaps I made a mistake in not inviting their representatives. The discussions at this meeting were long and varied and all were recorded by a court stenographer and published as a pamphlet.

At this meeting I expressed my astonishment that approximately one million dollars was required to equip a plant which was a brand new facility and which had been in operation only six months. I could not understand how the original operator, Remington-Rand, could have been expected to produce satisfactory equipment in an uncompleted plant, if the plant were in fact uncompleted as seemed to have been the case. This matter has never been explained to my satisfaction.

At any rate, my "town meeting" pamphlet served as the point of departure for a Congressional Investigation which, I learned later, inquired into the hiring by the Managing Agent of a firm of management experts, one member of which was a reserve officer on duty in the Bureau of Ordnance.

My instructions with respect to inventory stated as follows: "Under the circumstances it is not deemed necessary for you to take full and complete inventory of 'The Plant' and the work in process unless after taking possession you deem it advisable. The question of inventory when the property is returned to private operation is reserved for future determination. . . ."

The main building, including all facilities therein, as well as all raw material, was the property of the Government. The land and certain minor facilities were the property of Remington-Rand. The area occupied by this activity had always been under the control of guards. For this reason and because of assurances that the plant would shortly be operated by the Norden Company under contract with the Bureau of Ordnance, and that the Executive Order would be shortly withdrawn, no inventory was taken of the plant at the time of seizure. At that time, it is estimated that there was one million dollars' worth of spoiled material on hand but part of this material was eventually salvaged and worked into sights produced during the period of occupation. On 3 July it was estimated that there was four and a half million dollars' worth of material in the process of manufacture.

While figures have been given in regard to the reduction of the cost of bombsights, no cost figures on material produced by this plant can be regarded as final without a word of qualification. As regards the alleged cost of bombsights during the period of occupancy, these figures are obviously lower than they would have been had any means existed of adding to these costs the cost of material salvaged from the accumulation of spoiled material, estimated by some at an original value of one million.

Operating costs from the inception under both the management of Remington-Rand, Inc., and Carl L. Norden, Inc., of the plant at 1051 South Main Street, Elmira, New York, had been measured on the concept of cash disbursed. A further observation which makes for a unique cost concept is the extraordinary skill required for the manufacture of precision instruments and the comparatively short period of actual satisfactory production. Both concepts encourage the adoption of averages by months or some other arbitrary period for determining unit costs. Unless the period of production is sufficiently extended to permit the allocation of costs on the basis of a normal cycle of operation conditions, cost figures contain many non-repetitive items. For example, training costs incurred by Remington-Rand, Inc., are only one portion of the ultimate cost of acquiring the "know-how." There are the rework costs and cost of parts scrapped which must be considered. In addition, certain costs were incurred for tools deemed necessary for adequate process inspection. Under the circumstances, it is correct to say that at no time during the operation of this plant, either by Remington-Rand, Inc., or by the Navy, were any comparable unit costs compiled representative of stabilized operations.

This matter of costs has been brought to attention because it is obvious that commonly accepted methods of cost accounting do not seem to have been practiced in the production of bombsights. At any rate, whatever the actual costs were, bombsight production was raised from 153, for the period from June to November 1943, to 2,672, for the period from December 1943 to July 1944, under Navy management. Thus the purpose of the seizure, namely to obtain increased production from this plant, was attained by July 1944.

Sometime during May 1944, I learned from the public press of the employment by the managing agent of the firm of Corrigan, Osborne and Wells, management experts. This firm had been employed without my knowledge or the knowledge of my Deputy, Commander Harrison. This episode was the subject of Grand Jury and Court action following the Congressional Investigation, which resulted in at least one conviction.

By midsummer, operations of the plant had become routine, and on 3 July 1944 I was relieved as Office-in-Charge by Rear Admiral G. F. Hussey, Jr., Chief of the Bureau of Ordnance.

Seizure of the Plant of the Los Angeles Shipbuilding and Dry Dock Company

An order signed by the Secretary of the Navy on 27 September 1943 convened a special Naval Board of Investigation, of which I was the senior member, with authority to administer oaths and subpoena witnesses for the purpose of inquiry into the affairs of the Los Angeles Shipbuilding and Drydock Company. The plant of this company, located at San Pedro, California, was full of vessels building for the Navy and the greater part of the plant was owned by the United States government. Conditions at the plant were such as to arouse the apprehensions of the Chairman of the House Naval Affairs Committee, the Honorable Carl Vinson, and he was determined that the Navy Department should take action before the gross waste of public money at that yard became a public scandal. The contracts were of the cost plus fixed fee type.

The Board convened at San Pedro. Also present during the meetings of the Board was a sub-committee of the Naval Affairs Committee of the House consisting of the Honorable Jack Z. Anderson and the Honorable Warren G. Magnuson. The sub-committee participated in the proceedings of the Board. The report of the Board was approved by the Chairman of the Naval Affairs Committee of the House of

Representatives and by the members of the sub-committee. The Board found, among other things:

(a) That, taking into consideration extenuating circumstances, from $5,000,000 to $7,000,000 of the expenditures of the *Ajax*, a repair ship building for the Navy, was excessive and unjustified and that the cost of nine other vessels building for the Navy would be unreasonable, although not as expensive as the *Ajax*.

(b) That it would be unreasonable to believe that there would be any great reduction in the cost of construction in that yard until drastic methods were employed to produce a state of efficiency.

(c) That the yard was not then, and apparently never was, organized along modern industrial and shipbuilding lines; that the controls over production, flow of material, and expenditure of funds were inadequate or absent.

(d) That coordination between departments was lacking to a startling degree.

(e) That the yard lacked an adequate inventory.

(f) That principal events had repeatedly been over-optimistically scheduled, resulting in the excessive flow of materials into the yard, with consequent problems of storage and inventory.

(g) That production schedules, when existent, had not been adhered to, resulting in delays and high costs.

(h) That there never had been a sufficient number of competent key men in the management who were experienced in modern industry or shipbuilding; that the management conceded that the yard required the services of a capable general manager, but that while concurring it felt that more far-reaching changes were necessary.

(i) That constant changes in the key personnel prevented coordination between the various departments, had there existed any will for such coordination.

(j) That cost analysis and cost keeping for purposes other than for securing reimbursement from the government were entirely absent; and that this absence deprived the yard of its only means of furnishing an adequate check on the performance of superintendents, supervision, and direct labor.

(k) That no evidence of cost consciousness in the yard was found under the present management or any other; and that the fact could not be established that any cost conference had ever been held.

(l) That as of 30 August 1943, approximately $64,000,000 of public money had been expended at the yard for the construction of new vessels; and that those responsible for the expenditure of large sums had not been able to control them, nor had they been sufficiently impressed with their responsibility.

(m) That there had been and was an inordinate amount of idle time and loafing at the yard, which with other facts, indicated a lack of adequate supervision.

The Board recommended "that the Corporation obtain a staff of competent shipbuilders and adopt an organization satisfactory to the Bureau of Ships. Failing in this the following two alternatives are recommended which require in either event that the plant be taken over pursuant to Executive Order—

(a) The plant to be staffed with competent personnel furnished by the Navy Department and other shipbuilding plants.

(b) A management contract to operate the plant be executed with a competent organization."

After further negotiation between the Bureau of Ships and the Los Angeles Shipbuilding and Drydock Company, the Secretary of the Navy decided that the seizure of the plant of that company was necessary to prosecute properly the war effort and therefore at 10:00 a.m. on 8 December 1943, I seized the plant under Executive Order #9400, having been designated Officer-in-Charge, the Honorable Carl Vinson having insisted to the Navy Department that I be detailed to do the job.

I was accompanied at the time of the seizure by my staff consisting of Lieut. Comdr. Ethan A. Hitchcock of the Office of the Under Secretary of the Navy, General Counsel's Office, Lieut. Murray C. Hotchkiss, prospective Disbursing Officer, and Mr. Frank W. Marshall, Executive Accountant and Auditor, Bureau of Supplies and Accounts, Navy Department. We called at the office of Mr. Alfred F. Smith, President of the Los Angeles Shipbuilding & Drydock Corporation, to announce the seizure of the plant. Mr. Smith was absent and was represented by Mr. W. W. McComb, Assistant to the President. A copy of the Executive Order was delivered to Mr. McComb and at 10:30 a.m. printed copies of the Executive Order, notice of possession, and customary orders were posted throughout the plant and offices.

The corporation was instructed to close the books of account as of 8:00 a.m. on 8 December 1943. It was ordered that the wage payroll due and payable 10 December 1943 would be paid by the corporation.

Mr. McComb was informed that he, Mr. Alfred F. Smith, and the Comptroller, Mr. Beaman, were no longer on the payroll of the plant and that other severances would be announced from time to time. Shortly afterwards, the Public Relations Officer of the company was removed from the payroll.

At 12 o'clock, I made an address over the public address system to the employees of the plant, announcing the change in the management and the purposes of the Executive Order.

At 1:00 p.m., I called a press conference, and the representatives of the press were informed of the circumstances surrounding, and the purposes of, the seizure.

The Todd Shipyards Corporation, having been selected for managing agent in connection with the operation of the plant, the following members of the Todd Shipyards Corporation appeared at the plant at 10:00 a.m., 9 December, to discuss the task at hand and to make preparations for assuming control of operations under the supervision of the Officer-in-Charge:

J. D. Reilly, President, Todd Shipyards Corporation
J. Herbert Todd, Vice President, Todd Shipyards Corporation
Joseph Haag, Jr., Vice President, Todd Shipyards Corporation
Harry Hill, General Counsel, Todd Shipyards Corporation
J. R. Lamont, Director, Todd Shipyards Corporation
Arthur Sewall, Customer Relations, Todd Shipyards Corporation and other officials from various Todd Shipyards who were to remain at the plant of the Los Angeles Shipbuilding & Drydock Corporation to perform duties as indicated below.
F. D. Hesley, General Manager
J. J. Redington, Special Assistant to General Manager
Herman Minikine, Asst. General Manager for Repairs
R. W. Copeland, Asst. General Manager for New Construction
W. G. Abernethy, Chief Accountant
C. A. Lane, Administrative Assistant

I executed, as agent for the Navy Department, a letter of intent with the Todd Shipyards Corporation, to operate the plant. An initial and tentative organization chart for the operation of the plant was prepared by the Todd Shipyards Corporation.

On 10 December 1943 an agreement was signed with the American Appraisal Company to take a physical inventory of all materials and supplies in the plant of the Los Angeles Shipbuilding & Drydock Corporation as of 8:00 a.m., 8 December 1943.

A proposal from Peat, Marwick, Mitchell and Company to audit the balance sheet of the Los Angeles Shipbuilding & Drydock Corporation as of 8:00 a.m., 8 December 1943, was accepted and the accounting firm was directed to proceed.

On 14 December the inventory began.

On 15 December the audit began.

On 16 December the payroll account was opened in the Bank of America with an initial deposit of $450,000.

On 18 December 1943 at 1:15 p.m., the Honorable Jack Z. Anderson, U.S. Representative from the State of California, met with the Officer-in-Charge and his staff for a discussion of the seizure. Mr. Anderson together with the Honorable W. G. Magnuson, as will be remembered, constituted the members of a sub-committee from the Naval Affairs Committee of the House of Representatives which participated in the investigation of the affairs of the corporation in October of that year.

Legal Matters. The following action was taken, and procedures to be followed established, with respect to the business and legal arrangements for the plant of the Los Angeles Shipbuilding and Drydock Corporation.

Contractual relations with Todd Shipyards Corporation providing for its acting as my managing agent were initially established by Letter of Intent which was intended to be promptly superseded by a definitive agreement. It authorized Todd, and constituted Todd's agreement, to enter the plant with such personnel as might be required by me to perform such duties as I might assign; it indemnified Todd against any loss it might sustain from operations thereunder, unless such loss resulted from willful wrongdoing or gross negligence on the part of its personnel. The letter did not authorize Todd to enter into commitments on behalf of the United States and no provision was made for paying Todd for its services except in the event the Letter of Intent was terminated by me, whereupon, Todd would be entitled to such fair compensation as I might determine.

Because the definitive agreement was not concluded with Todd as promptly as contemplated, the Letter of Intent was amended as of 24 December 1943, so as to permit Todd to incur commitments in the name of the United States to the extent necessary for the operation of the plant, but subject to the following limitations: Commitments for services and materials required for Government work must have the prior approval of Supervisor of Shipbuilding or other appropriate Government representative; commitment for items ordinarily chargeable as capital expenditures involving a sum in excess of $1,000.00 must have my prior approval; commitments for overhead items other than usual and recurring overhead charges must have prior approval by the Disbursing Officer acting for me.

I authorized Todd to enter into commitments for the performance of Government and private repair work. Government work was to be performed in accordance with standard Government form for repair contracts. Private repair work was to be performed in accordance with terms established by me which, among other things, limited to $300,000.00 liability for loss or damage to a vessel being repaired. This liability was covered by insurance. The owners of vessels in the repair yard, other than Government vessels, were informed that work performed for them by this activity was performed subject to the terms established by me. The charges for private repair work, in accordance with industrial private repair work practices, were to be negotiated with the owner when the work was completed. The charges were to be based on standard commercial rates. Todd was authorized, as my representative, to negotiate and agree with owners as to the charges for work performed.

A proposed agreement was submitted to the Los Angeles Shipbuilding & Drydock Corporation providing for the proration of charges made for private work commenced and performed in part by the Corporation prior to the take-over, but completed by the new management after the take-over. Under the Government agency repair contract form, the corporation would simply be paid its time and material costs to the date of take-over and therefore no agreement for apportionment appeared necessary on Government agency work.

A new purchase order form, containing the required provisions for Government contracts, was prepared and substituted for the purchase order forms heretofore used by the corporation. This form was used in connection with all purchases made for the plant.

Instructions were given to the Purchasing Department and the Disbursing Officer that States Sales Tax would not be applicable to the purchases made by the present management and hence should not be paid. Todd was instructed to obtain exemption from Federal excise taxes in the manner prescribed in the Bureau of Ships memorandum on that subject.

While it was recognized that state laws applicable to the attachment and garnishment of employees' wages could not be enforced against the new management, it was nevertheless decided as a matter of policy to recognize such attachment and garnishment proceedings, and the Disbursing Officer was instructed to comply therewith.

As has been the custom in other cases of seizure, arrangements were made with the State Unemployment Insurance Commission pursuant to which the new management would make State Unemployment Insurance contributions to the State Fund on the same basis as did the corporation, and make the usual deduction from employees' earnings for such purpose.

The West Coast representative of the Navy Department Insurance Division made arrangements for insurance coverage formerly carried by the corporation to be amended and continued so that it would cover the interests of the corporation, and the Government, as their respective interests appeared, with Todd added as a third named assured under the majority of the policies.

A contract was entered into with the caterer who had been providing canteen and cafeteria service for the corporation. The agreement was in substance a license, permitting the caterer, without charge, and as an independent contractor, to operate canteens on the premises. The term of the agreement was six months with right of either party to terminate on thirty days notice.

The question of rental payable under the lease between the Los Angeles Board of Harbor Commissioners and the corporation was discussed with a representative of the Harbor Commissioners and a representative of the Bureau of Yards and Docks West Coast Real Estate Division. It was indicated that the Board would probably be willing to have the Navy Department continue under the existing corporation lease, with payment of rental based upon a percentage of receipts from repair work plus expenditures on construction work. The percentage would be the same as that which was provided under the lease with the corporation, where gross receipts was the figure to which the percentage was applied. It was also indicated that willingness existed to agree upon a flat annual rental if the Navy Department preferred that form of payment. Rental for 1943 would be paid proportionally by the corporation and the Navy Department and final arrangements for 1944 rental would be made at some time in the future.

Notices of termination of vessel construction contracts, Navy, NOd-1577, NOd-1857, NObs-888 and NObs-934 were served upon the corporation on 8 December 1943. These contracts constituted all the construction contracts held by the corporation, with the exception of NOd-1515 which provided for construction of the *Ajax*. That vessel had been accepted by the Department prior to 8 December, and it was felt no advantage could result from termination of the contract therefor.

The corporation had established a line of credit with The Security-First National Bank of Los Angeles and the Bank of America. Loans effected thereunder were secured by the Department's V-loan guarantee to the bank to the extent of 50% of the amount loaned, and by the assignment to the bank of sums payable to the corporation under the vessel construction contracts. Shortly after the date of take-over, and termination of vessel contracts, the corporation needed funds for meeting its payroll. The bank refused to extend further credit without assurance that the Department would continue reimbursement to the corporation of allowable costs incurred by the corporation under the terminated vessel contracts. I gave the necessary written assurance to the bank, reserving the right however, to withhold such part of payments as might in the Department's opinion be necessary to insure no overpayment by the Government under the provisions of the vessel contracts as terminated. It was understood that the corporation as of 4 January had paid off all but $1,000.00 of its loan from the bank. Consideration was given to the advisability of the Department's terminating its V-loan guarantee.

Audit of Books and Records. In taking over a business such as the Los Angeles Shipbuilding and Drydock Corporation, the establishment of the value of assets and liabilities became a major task because of the numerous and complex transactions with respect to taxes, plant values, depreciation, commitments for materials, and other items. Moreover, since the books and records of the corporation were continued during the occupation period, the soundness of the prevailing accounting principles necessitated a review by independent accountants to establish the validity of the opening balance. In other words, it was most essential that an accurate determination be made of the financial status of the corporation at the date of seizure (8 December 1943). By establishing such status, complete and accurate accountability was maintained throughout the period of occupation, and thus numerous claims were avoided with respect to the values of the properties and liabilities related thereto. It was in order to carry out this work expeditiously that Peat, Marwick, Mitchell and Company, Public Accountants, were engaged for this purpose on the day of the seizure.

Cost Accounting. There was no vestige of a cost system in the plant. The so-called accounting department was only used to see that the contractor got his payments from the Government. In determining costs by vessels the practice had been to accumulate the amounts paid for wages, materials as received (regardless of when installed) and the general expense for each vessel, without considering the adequacy of the supervision effectiveness of the labor load by trades, or the timeliness of material fabrication. Such cost keeping as previously indicated is simply a method for validating expenditures in connection with reimbursement.

One of the first tasks was to set up a job order cost system which would reflect the labor costs by jobs, by trades, and by vessels. At the beginning it was not possible to locate material charges thus because that could not be done until the inventory had been completed. Also, to locate material charges correctly it was necessary to break up the vicious practice of charging material directly against a ship as soon as the material arrived in the yard, whether or not the material was ever incorporated in the ship.

This task fell upon Mr. Marshall and myself. Neither one of us had ever set up a job order cost system and we succeeded surprisingly well. As a matter of fact, I never heard of any shipyard before having a job order cost system for new construction—for repair, yes.

For the purpose of building a ship, it is necessary to divide the ship into a number of chunks or bundles structurally speaking and into a number of systems (propelling, light and power, interior communication, fresh water, etc.) and subdivisions thereof in order to procure material and do work according to plan. Also the ship must be regarded as an aggregate of compartments and all procedures must have due regard to compartmentation. This is known as compartment scheduling.

Fortunately for setting up a cost system, the work of subdividing a ship into structural chunks and mechanical and electrical sub-divisions had already been done for Hull #61—*Norton Sound*, aviation tender, one of the later contracts, and for subsequent vessels. These chunks and subdivisions were given a job order number, a condensed title and a budget or estimate in man hours of direct labor to complete. The accounting division kept cost to date on each job order in dollars and cents and converted estimated man hours for each job order into dollars and cents by using a conversion factor that is a relationship between man hours and average wages per hour. At routine intervals, the Assistant Manager for New Construction had placed on his desk a statement showing the estimated cost, and cost to date, of direct labor for each existing job order.

This cost keeping could not be done on the *Hector* (Hull #59) which was delivered 15 February 1944, nor on the *Jason* (Hull #60) which was to be delivered 30 April 1944, since both were too far along.

Shop expense was not budgeted and an investigation was made to ascertain whether or not such a system could be established easily and to what extent it was warranted.

All overhead charges had to be investigated to discover waste and to remove undesirable or unnecessary individuals from overhead expense. Overhead expense was excessive.

From the cost data assembled, a periodic report was prepared for my information as well as for the new management, showing the progress by shops and the labor costs to achieve such progress. Reports were presented currently, and not on a historical basis so that proper steps could be taken to correct shop practices which were indicated as too costly and were causing delay in construction.

A highly important function of a plant is the prompt payment of payrolls. It was necessary for new checks to be printed and a signature plate secured, all of which had to be on hand by 15 December 1943, to pay the roll for the week ending 12 December 1943. The equipment needed for payroll purposes was received and the workers paid as scheduled on 17 December 1943. All this was preceded of course by arrangements with the banks so that the necessary funds would be available.

Disbursing under Navy procedure for materials procured differed from that followed by the corporation because the corporation ordinarily paid for materials on discount or due date regardless of the fact that materials had not been received.

Inventory Control. One of the outstanding conditions which continuously confronted the corporation was the inadequacy of inventory control of raw materials. This condition resulted in materials not being procured in accordance with ship construction schedules which resulted in confusion in the plant and warehouses. In order that materials utilized in the construction of vessels could be properly controlled and costed on the basis of installation on a particular vessel instead of on a procurement basis, an adequate segregation by categories and value was essential. Such a control could only be effectively established through an accurate starting inventory. Rather than impede production by diverting employees from their regular duties, and because of the enormousness of the task, I had engaged the American Appraisal Company to make a complete inventory of the plant.

This inventory became the basis for establishing a material control which resulted in:

(a) Increased efficiency in scheduling
(b) Determination of excess quantities
(c) Proper allocation and storage
(d) Material values at 8 December 1943
(e) Current cost reports of vessels under construction

A complete control, accurately established, and currently maintained was also of inestimable value in the matter of procurement. This official count, independently prepared, was tabulated in such form that each department had access thereto in connection with their respective responsibilities in operating the plant.

Labor Relations. Prior to 8 December 1943 this plant was operated under a contract between the Los Angeles Shipbuilding & Drydock Corporation and Local #9 CIO, of IUMSWA. Under this contract an employee after 30 days was given the opportunity to join the union and to sign an authorization for the corporation to deduct from his wages an initiation fee and monthly dues (check off). If he joined the union, the corporation was bound to discharge him if the union certified that he was no longer in good standing (union maintenance). This was not a closed shop as there were employees who did not belong to the union. It was alleged that about 800 were delinquent in their dues. Only employees in journeymen, helper, and laborer ratings were unionized.

Labor Relations were conducted by me on the following premises:

(a) The Navy Department does not recognize unions.
(b) Under the Executive Order, the Plant being Government operated, the employees are Government employees.
(c) The Navy Department does not engage in Collective Bargaining.
(d) The Navy Department uses as a guide any agreements entered into by the former management and the local union as long as such agreements do not interfere with the objectives of the Executive Order.
(e) On the other hand, the Navy Department will not enter into any agreements with the union.
(f) The usual channels for handling grievances are observed but there is no arbitration. Major grievances must be presented to the Officer-in-Charge, whose decisions are final. All grievances are settled on their merits.
(g) The check off is paid where employees have signed an agreement for that purpose with the union.
(h) Employees are not coerced by the management to pay dues.
(i) New employees are not coerced by the management to join the union.

Upon arrival at the plant, both the Todd management and I repeatedly exhorted the foremen, supervisors, and wage earners to stop the loafing at the plant. Very little was accomplished by these efforts.

The day before Christmas many employees came to the plant, checked in, and checked out. More checked out at noon. Friday afternoon, conditions were most unsatisfactory. Very few employees were working aboard ship. Most of them were engaged in group conversation and many were visiting. No attempt was made to dissemble. They had simply knocked off and were waiting for the whistle to blow to collect time for doing nothing. Evidence of moderate drinking was widespread. A certain number of drunks were sent home by their foremen.

As a result of this situation, an order was issued that any employee would be discharged for loafing, quitting before whistle blow, intoxication, drinking in the yard, or having liquor in his possession. Employees were discharged for some of these reasons. The local union protested this order and requested that three days' suspension be substituted for discharge. I refused to change the penalty on the grounds that both labor and the former management individually and collectively had made a dismal failure of controlling loafing at this plant and that they could both step aside and let someone else try it.

Labor was told that an attempt was being made to eliminate the large number of unfit from the payroll, and that it and former management jointly shared the responsibility for waste of public money in a plant where it is alleged as many as 1,000 employees had been asleep on ships at night.

Labor agreed to abide by the discharge order and expressed every desire to cooperate in straightening out the plant.

The union later protested the subcontracting of work to outside concerns who happened to employ A.F.L. labor. The union was informed that work would be subcontracted whenever necessary to achieve the purpose of the Executive Order and that as soon as plant labor could handle the work in hand, no more subcontracts would be let.

It was necessary to divide the Personnel Department into two major subdivisions: (1) Labor Relations and (2) Administrative Personnel. A few of the chaotic conditions which existed in this Department may well be enumerated.

The shop stewards were running wild and seriously interfering with the foremen in the execution of their duties. Shop stewards would tell the foremen whom to place on particular jobs. Shop stewards in one department would interfere with the operations of other departments. Foremen had not been instructed in their duties. They were never taken into the confidence of management with the result that they did not know how to handle labor problems or run their shops.

It was not long before these conditions were corrected through foremen conferences and an educational program. Conferences with union representatives disseminated the information throughout the yard that shop stewards would not interfere with production and that they would deliver a day's work for a day's wage.

The handling of grievances was expedited.

A uniform discharge policy was put into effect.

General Organization. Further acquaintance with this plant served to corroborate the conclusions of the Naval Board of Investigation that the organization of the plant was unsatisfactory fundamentally. Reorganization proceeded with the idea of establishing definite controls over planning and scheduling, material, production, and cost. The planning section was made an executive function of management, that is, its orders were made mandatory, rather than permissive or advisory. It was given control of the flow of material from the bills of material stage to the delivery of material to the point at which it was fabricated. The technical section was subordinated to the planning section instead of vice versa. As brought out in the investigation, the plant had no top control organization or "front office." This was corrected and a small authoritative group set up in the former executive offices consisting of the Manager, Special Assistant to the Manager, Assistant Manager for Personnel, and Comptroller. Arrangements were made with Gibbs & Cox, 1 Broadway, New York, New York, to furnish engineering information beyond the capability of the yard staff.

A better distribution of duties between the Hull Superintendent and the Outfitting Superintendent was effected and both became subordinate to the General Superintendent.

Weak and inefficient leadership fails to handle routine

matters in time with the result that they finally have to be handled as urgent. The new management was immediately confronted with such problems.

There was no routine method of evacuating sick from the plant, and such a method was established.

In some places the plant was a vast mud puddle. Railroad trackage was not ballasted and in one instance the track was so depressed that the contents of a flat car were dumped into the mud. Movements of pedestrians were interfered with by mud. This condition was corrected.

No official of the plant was responsible for the stability of ships after launching and no one in the plant knew what their condition was in this respect after launching. This condition was immediately corrected.

The cafeteria system was most unsatisfactory and had been so for many months and nothing had been done about it. A survey was instituted and recommendations and estimates submitted to the Bureau of Ships to correct this situation.

An inspection of the floating drydock indicated that it might be unsafe to operate. At the request of the management, Todd arranged for a survey by the firm of Frederic R. Harris, Inc.

It was alleged that the welding machine situation had been unsatisfactory for months, the old company having failed to convince the Supervisor of Shipbuilding of the necessity for more welding machines, so the new management stepped into an old controversy. All arguments pro and con were carefully considered and as a result 288 welding machines were promptly added to the plant equipment.

Necessity for an Executive Order. Conditions at this plant were found to be even worse than reported by the Naval Board of Investigation. It would not have been possible to correct these conditions by any measures short of an Executive Order, so long as the management which was responsible for the existence of such conditions remained in control.

The Executive Order directed the Secretary of the Navy

to seize and operate the plant in order effectively to construct, convert, and repair vessels required by the United States. It clothed the Secretary with unusual and extraordinary power to cope with the emergency created by the plant's failure to produce satisfactorily. The Executive Order was resorted to only after all other methods had failed to produce the results urgently required for the war effort. The Secretary designated an Officer-in-Charge as his representative to exercise the emergency powers vested in him and gave the Officer-in-Charge instructions as to the procedure to be followed in carrying out the objectives of the Executive Order. Under the authority of these instructions, the Officer-in-Charge retained Todd Shipyards Corporation to act as the managing agent of the Officer-in-Charge and to manage the plant in such capacity. The Officer-in-Charge remained in command as long as the Executive Order remained in force, and in spite of the designation of the managing agent, the Officer-in-Charge determined all matters of policy relating to the operation of the plant, its labor policies, and its financial affairs. If the Officer-in-Charge became dissatisfied with the management chosen, he had the authority to terminate the services of the managing agent and provide other management.

The absolute authority of the Officer-in-Charge was vital to the success of undertakings under an Executive Order and was as much justified as the Executive Order. The conditions which will be encountered on the seizure of a plant can almost never be predicted. The Officer-in-Charge must therefore be given broad authority to act promptly and on the spot if he is to assume responsibility for holding together such "going concern" qualities as may exist at the plant at the time of seizure. It follows that his sole and direct responsibility should be to the Secretary of the Navy, with reliance being placed on the discretion and judgment of the Officer-in-Charge for determining to what extent the purposes of the Executive Order will permit the following of usual and routine procedures. An Executive Order for seizure and operation of a plant is only issued in cases where the normal relations between a Bureau and a company have broken down and where corrective measures must be applied which are far beyond the authority of a Bureau. The issuance of such an order is definitely not a reflection on a Bureau, but the Bureau must withdraw from the situation, or at least subordinate itself in respect to the company, to the Officer-in-Charge. The office of the Officer-in-Charge is an extension of the office of the Secretary of the Navy and, as such, is senior to any Bureau and even senior to the office of the Chief of Naval Operations. Any Bureau inspectors should report to the Officer-in-Charge for additional duty. The importance of single responsibility may be best illustrated by showing the number of conflicting authorities to whose control I would have been subject in operating the plant of Los Angeles Shipbuilding & Drydock Corporation, had I not been under direct orders from the Secretary:

Commandant, 11th Naval District
Industrial Manager, 11th Naval District
Supervisor of Shipbuilding, 11th Naval District
Commandant, Naval Operating Base, Terminal Island, San Pedro
Assistant Industrial Manager, 11th Naval District
Public Relations, 11th Naval District
District Labor Relations, 11th Naval District
Assistant Labor Relations, 11th Naval District
Liaison with War Manpower Commission, 11th Naval District
Office of War Manpower Commission, Los Angeles
Regional Office, Department of Labor, Los Angeles
Regional Office, War Manpower Commission, San Francisco
Assistant Coordinator Ship Conversion and Repair, San Francisco
Special Representative of Ship Conversion and Repair, San Francisco
Regional Compliance Manager, War Production Board, Regional Office, San Francisco

The wishes and requests and orders of all agencies were complied with and the intent of all directives and decisions were adhered to as long as they did not interfere with the emergency purposes for which the Executive Order was issued; namely, the speeding up of production, and elimination of scandalous waste of public money at this plant.

Shortly after the seizure, complaints were received from the War Production Board, that the corporation had violated the Controlled Material Plan and was to be cited for court action; the Bureau of Internal Revenue agents called and cited the corporation for violations of laws in regard to salaries; agents of the Department of Labor called and charged violations by the corporation of the Child Labor Act; other agents of the Department of Labor called and charged violations of the 8-hour law and representatives from the State of California called and complained of violations of the State Sanitation Laws.

So many complaints, so soon after the seizure, for acts committed during the period of the former management, might suggest more than a coincidence.

Progress and Operations. By 22 May 1944 I was able to report considerable improvement in conditions at the plant and considerable progress in reorganization. Of course, reorganizing a shipyard without slowing down shipbuilding is not the easiest thing in the world, but improvements already made were so fundamental that they were responsible for a net improvement.

In production the main objective was to increase the volume of work per individual by increasing the effectiveness of direct supervision, by more careful planning, by elimination of habitual non-producing employees, by increasing emphasis on training, and by improving safety.

As a result of the audit of amounts due vendors, liabilities were increased from the original list furnished by the company from $838,855.83 to $2,274,476.31 and the investigation was not completed.

Considerable progress had been made in overhauling overhead expense and on the inventories of materials. For the first time in the history of the yard a man-hour budget of total direct hours was estimated which would be required to complete the vessels to be built in the yard. The repair yard showed a rather large profit.

I do not relish telling the story of an uncooperative union. My reputation for being fair—and not anti-union—was responsible for my going to San Francisco, to break Lodge 68 A.F.L. with the open and public backing of such A.F.L. leaders as William C. Frey and Harvey Brown. Brown was a recent acquaintance but Frey I had known for years and always had respected him as a temperate, frank advocate of the rights of labor within our American concept. Frey is a great patriot.

The thing has to be done, however, in order to record faithfully, not so much my experience in developing the powers under an Executive Order but the attitude of some people in the labor unions.

The use of brute force on my part in solving my difficulties with this local C.I.O. union was fortunately averted by the great understanding and support of the President of the C.I.O. Maritime Workers, Mr. John Green of Camden, New Jersey. Mr. Green and I had mutual confidence and respect for one another, and I am indebted to him for his wise and considerate support. We had been in contact with one another for a great many years.

While I was attempting to operate and reorganize Los Angeles Ship with the help of Todd Shipyards Inc., I was also the Officer-in-Charge of the "N" Plant, Remington-Rand, Elmira, New York. Where to be was always a problem. I happened to be in my main office in the Navy Department, Washington, D.C., when I received a telegram from Walter S. Pollard, Senior Administrative Director of Local 9 C.I.O., San Pedro, under date of 8 April 1944.

According to him, everything had gone to pot. The Manager of Los Angeles Ship, Mr. Herbert Hesley, had issued an order that employees leaving a ship for any reason whatsoever other than at quitting time must check out upon leaving and check in upon return. We were charged with "fascist regimentation."

I have spent many years of my life with shipbuilders and their employees. This union was the most inefficient aggregation I have ever seen assembled in all my 46 years in the Navy. The loafing was scandalous and most difficult to control because of war expansion. The country had to make supervisors out of men who in many cases weren't really first-class mechanics and first-class mechanics were made of employees who could barely master one small part of their trade. Under these conditions we should understand the predicament of some of the really fine labor leaders in this country who suddenly found some of their local unions expanded several hundred times in membership without any time at all to develop competent leaders.

In reply to Mr. Pollard's telegram, I informed him that my managing agent was authorized to entertain any constructive suggestions from him as to a better way to eliminate waste of public funds and to expedite construction.

The fine relationship which had existed between the union and the company management would have been the same or similar had the union and the company deliberately agreed to leave one another alone. I do not state that collusion existed; I reiterate however that the same results could have been achieved by agreement.

Let us recall that in my telegram of 10 April 1944 I had informed Mr. Pollard that if the union had a better way to stop the scandalous loafing at Los Angeles Ship, I would consider favorably such a proposition. In spite of this, on 13 April 1944, three days after my conciliatory telegram to Mr. Pollard, there was a ruckus in the yard. How I wish I could have been there!

Mr. Walter S. Pollard, Senior Administrative Director of C.I.O. Local 9 had *directed* through the Chief Shop Steward and others that the check-in, check-out should be disobeyed. Thus did this union deliberately attempt to sabotage our attempts to stop the scandalous waste of public money at this yard—the United States at that time being in a state of war. Thus did the union reply to my attempt to get from them a substitute and better measure.

The Chief Steward later acknowledged that he had not been informed by Mr. Pollard prior to this union's unwarranted, illegal, and unpatriotic interference with the Navy management of this yard of the contents of my telegram to Mr. Pollard.

But to describe the ruckus, my reports indicate that there were 400-500 employees standing around idle, on government time, discussing the suspension of a Shop Steward for violating the check-in, check-out order of 4 March 1944. Mr. Walter Pollard appeared on the scene and, so I am told, gave a speech to the effect that check-in, check-out was a direct form of fascism and if necessary he would go directly to Washington and inform the President of the United States about this order. With a final heroic effort, Mr. Pollard lifted his hat and told the employees that their rights would in no way be molested, and that if necessary the Truman Committee would be notified of Navy management's way of handling personnel.

The *Ajax* cost $24,806,821 while her sister ship, the *Vulcan*, cost $12,873,090, and the joint Congressional Naval Investigating Committee disclosed the disappearance of $5,000,000 to $7,000,000 on the *Ajax*. No one ever assigned this loss of public money to fraud. They did assign it to an inefficient management and a poorly-organized labor union.

Why speculate about the size of our National Debt?

On 15 March 1944 I presided at a protracted hearing between representatives of the union and my managing agent (Todd) relating to grievances which had gone through the first four steps of the grievance procedure. After this long and important grievance session, I made my announcement in respect to what would be allowed. I always maintained during my various tours of duty in many cities, in many plants, that I did not arbitrate—I dispensed industrial justice. The Government is the sovereign power. It does not bargain.

Matters came to a head on 13 May 1944.

I had steadfastly refused to recognize any contracts in force between union and management at the time of government seizure.

I think the report of a conference between me and Mr. Pollard and Mr. Shapiro of Local 9 is a masterpiece of brevity.

Present: Messrs. Pollard, Shapiro, Redington, Hawes, and Obst, myself and Lieut. M. O. Damon (Counsel).

"*Admiral Bowen*—Proposition number one.

"*Mr. Pollard.* We happen to be your guests. I do not know why you invited me over here.

"*Admiral Bowen.* I didn't invite you. I heard you wanted to see me.

"*Mr. Pollard.* We do not want to discuss classification. Mr. Winstead will discuss that on Monday. The time has come when Local 9 should start taking care of Local 9 and that means one thing—that our contract should be put into full effect immediately.

"*Admiral Bowen.* It will not be put into effect immediately.

"*Mr. Pollard.* Then our business is finished."

Note: Mr. Pollard and Mr. Shapiro then withdrew from the meeting without further comment.

Mr. Louis L. Livingston, Department of Labor Conciliator, called upon me on 19 May 1944 for the purpose of arranging a conference among me, my management, and the Labor Union in regard to the case of three employees discharged for cause. Mr. Livingston exhibited a telegram from an official of the Department of Labor directing that such a meeting be held and that, failing agreement, the contract called for arbitration. I informed Mr. Livingston that operating under an Executive Order, the question of the three dischargees would be settled by me finally—period. That he or no one else could tell an Officer-in-Charge to hold a conference with a labor union or anyone else. That under seizure, there was no arbitration. Mr. Livingston stated that under the circumstances he could no nothing but report my refusal to follow the grievance procedures of the contract (which I never recognized) to the Secretary of Labor, and that the Union would probably refer its contention that I was bound by the contract (same contract which I never recognized) to the National War Labor Board or the President.

Mr. Livingston never took the trouble to obtain a copy of the Executive Order, under whose authority I seized the plant. He was not aware that it was posted in the plant. He expressed surprise when handed a copy to find that the Secretary of the Navy ". . . may select and hire such employees . . . and . . . may exercise any existing or other contractual or other rights of said Corporation. . . ."

I further warned this representative of the Department of Labor that it would be a pious idea to keep Executive Departments of the government from fighting and not to put such a fight up to a busy President.

On 14 May 1944, Mr. David L. Bingham, field representative of the National Labor Relations Board, appeared before me to argue the suspension of two guards for cause. He claimed their suspension was discriminatory and in violation of the National Labor Relations Act. I explained that all those things had nothing to do with an Officer-in-Charge under an Executive Order. He further asked me to post a notice in the yard stating in substance that I would desist from unfair labor practices in the nature of discriminatory suspensions. Of course I did no such foolishness and no one ever made me do it.

The War Manpower Commission was also a great nuisance until the Honorable Paul McNutt decided a case in my favor.

A few weeks after the seizure, I fired a former official of the company for cause. I received a summons from the Los Angeles War Manpower Commission Agency to attend a hearing on the firing. I did not attend and allowed no one of my representatives to attend. The power of an Executive Order was being challenged. I received a summons to attend a similar hearing at a higher War Manpower Commission in San Francisco. I did not obey the summons. At some stage of the game I was ordered by the War Manpower Commission to reinstate the individual and pay him back pay. I did not. Finally, the case went to Washington and Commissioner McNutt decided in my favor.

During this period of close-in fighting with the union and many branches of government, I was particularly fortunate at having at the other end of the long-distance line one Charles F. Detmar, an astute taciturn lawyer from New York, who was in Secretary Forrestal's office. I faithfully reported everything to him, after I had done it, and he always approved. I doubt very much that he ever reported anything I did to Forrestal. The reason is obvious.

By 10 July 1944 operations at the plant had become so far standardized that I was relieved as Officer-in-Charge by the Chief of the Bureau of Ships, with Todd Shipyards continuing as the Bureau's agent. Based on Todd's detailed reports, it seems clear that the seizure and operation of this yard saved the government in excess of $13,000,000.

The Todd management was quick to see the advantages of the costing system set up by Mr. Frank W. Marshall and myself. They agreed with me that accounting is the most powerful tool available to an industrial manager and, afterwards, Mr. John D. Reilly, President, and Mr. Joseph Haag, Jr., Vice President, brought all Todd Shipyards in line in this respect with Los Angeles Ship, which was probably the first commercial shipyard engaged in new construction and repair which was completely budgeted.

Great credit for the success in reorganizing Los Angeles Ship is due to Fred D. Hesley, General Manager, and Joseph J. Redington, his special assistant. They were a natural team. Redington was a retired Lieutenant Commander in the Navy. He and I took our first cruise together on the armored cruiser, *Maryland*, in 1905-1906. Great credit is also due to Mr. Frank W. Marshall, Executive Accountant and Auditor of my staff, who, as previously stated, helped me set up a job order cost system and also made many contributions toward straightening out this mess.

Seizure of the Bridgeport, Connecticut, Plant of Jenkins Brothers, Inc.

On 14 April 1944 at 4:15 p.m. my staff and I took possession of the plant and facilities of Jenkins Brothers, Inc., Bridgeport, Connecticut. Mr. James L. Dunn, Assistant General Manager, senior ranking company official present, was advised that the plant had been seized pursuant to Executive Order #9435 dated 13 April 1944 and he was furnished with copies of the order, the letter of the President, and the notices of seizure which constituted Orders No. 1 and No. 2.

The seizure of this plant illustrates another example where the Navy financed the operation of a plant during the seizure. In fact, the whole operation at this plant was financial as no other problems were involved. In order to assure the collection of all cash receipts contributable to the period of Navy operations, possession of all branch offices was also taken with the understanding that all collections made on invoices dated 15 April 1944 or later must be remitted to the Officer-in-Charge. All collections and income received from the branch offices were paid over to the Navy Controller at Jenkins Brothers, Inc., 80 White Street, New York, N.Y.

For the purpose of continuing insofar as possible the normal administration and fiscal routines of the corporation during the period of the occupation of the plant pursuant to Executive Order, I promulgated an order which is included at the end of this section. This order was devised by my fiscal consultant, Mr. Norwood P. Cassidy, and is included here with the hope that it may be useful to some other Officer-In-Charge in some future seizure under similar circumstances.

The financing of the plant by federal money was authorized by the Secretary of the Navy in the allotment of $2 million from the appropriation "17x0300 Naval Emergency Fund."

As of 9 June 1944, an agreement was executed by Mr. A. J. Yardley, Executive Vice President of the Corporation, and Mr. John P. Yarasawich, President of Jenkins Valve Local No. 6123, and Irving Richter, International representative for the Union whereby the provisions of the National War Labor Board were accepted by the parties as amendment to the collective bargaining agreement signed 13 January 1944. The National War Labor Board unanimously voted that the execution of this supplemental agreement constituted compliance with the directive order of their board.

As of 14 June 1944, the Attorney General of the United States approved the proposal of the Secretary of the Navy for the compromise and settlement of the rights and liabilities existing between the United States and the corporation as the result of operation pursuant to Executive Order. As of 15 June 1944 the corporation accepted the proposal of the Secretary of the Navy as a basis of settlement of the respective interests of Jenkins Brothers, Inc. and the United States Government in connection with the possession and operation of the plant pursuant to Executive Order.

Government possession, control and operation of the plant and facilities of the corporation was ordered terminated by the Secretary of the Navy as of 12:01 p.m. on 15 June 1944, upon determination that the productive efficiency of the plants and facilities had been restored to the level prevailing prior to the threatened interruption of production referred to in the Executive Order.

In spite of all these fine words, the Navy was drawn into seizing and operating the plant of the Jenkins Brothers because that corporation would not comply with an order of the National Labor Relations Board directing them to pay back-pay to their employees because of a belated decision of the National Labor Relations Board. In my opinion, the Navy had no business at all participating in a procedure which was designed to enforce a decision of the National Labor Relations Board and that had nothing whatsoever to do in fact with cessation of or unsatisfactory production. As a matter of fact, no complaint had been made in respect to production.

The seizure and operation of this plant was facilitated by the fact that the plant, main office and branches were all connected by teletype.

The order devised by Mr. Cassidy was as follows:

Administrative and Fiscal Procedure Used in the Seizure & Operation of the Jenkins Brothers Plant

This will confirm and correlate financial arrangements outlined in the exchange of teletype messages between the Officer-in-Charge and Jenkins Bros. on April 19, 20, and 21st, 1944.

The Officer-in-Charge will meet all current operating expenses of the corporation attributable to the period of Navy management commencing at 12:01 a.m. April 15, 1944 through reimbursement to the New Jersey Corporation at its home office, 80 White Street, New York, N.Y., for all applicable cash disbursements made by the New Jersey Corporation attributable to such period. The Officer-in-Charge reserves the right, however, to deny reimbursement for certain expenditures which, although proper under corporate operations, would not be items properly payable by the Navy Management. Expenditures as to which such right is reserved include, but are not limited to, dividend payments, legal expenses, State and Federal taxes measured by income, expenditures which under generally accepted accounting practices would be charged to capital account, and charges for depreciation. The extent of reimbursement will be subject to the sole determination of the Officer-in-Charge.

All income and receipts from operations, whether received by the New Jersey Corporation or the New York Corporation, arising out of shipments or deliveries from inventories made after 12:01 a.m. April 15, 1944 or from other transactions attributable to the period of Navy Management which commenced at that time, is to be paid over to the Officer-in-Charge through the home office of the New Jersey corporation. The Navy Comptroller, Jenkins Bros. home office, will maintain a complete record of all expenditures made by and income received by, the Officer-in-Charge. This record will reflect currently the surplus or deficiency of such income in relation to such expenditures.

By these arrangements it is intended to put into effect a simple financial plan, paralleling as far as possible the practices of the New Jersey and New York corporations in effect prior to April 15, 1944. While further details developing in practice may require subsequent consideration and determination by the Officer-in-Charge, the following is stated as indicating the most important elements in the plan:

(a) For purposes of the plan, the New York Corporation is to be treated as a branch office, and all references to Branch offices will be deemed to include the New York corporation unless otherwise specifically stated.

(b) The New Jersey corporation shall make cash advances to all branch offices and the Bridgeport plants, for the payment of current operating expense, and shall pay its own current operating expense. Such cash advances and payments shall be administered by Jenkins Bros. personnel heretofore or hereinafter designated by Jenkins Bros. and shall be subject to the procedures and restrictions followed and imposed by Jenkins Bros. prior to the period of Navy management.

(c) Reports of cash disbursements applicable to such advances to branch offices and the Bridgeport plants shall be made to the home office by Jenkins Bros. personnel in the usual manner. In addition, reports of such disbursements shall be made by representatives of Jenkins Bros. primarily responsible therefor, to the Navy representative of the Officer-in-Charge. Branch office managers, with the exception of the Bridgeport branch office manager, will make such reports to the cognizant supervisory Cost Inspector; the reports of the Bridgeport plants and the Bridgeport branch office shall be made to the Navy Comptroller at such plants; the reports of the New York corporation shall be made to the Supervisory Cost Inspector; 21 West St., New York, N.Y. The New Jersey corporation shall also make similar reports of its cash disbursements for operating expense to the Supervisory Cost Inspector at New York. All such reports shall be certified by the responsible Jenkins Bros. personnel to the effect that all expenditures included are proper charges to Jenkins Bros. and that they have been made pursuant to authority granted by Jenkins Bros. The specified Navy representatives of the Officer-in-Charge shall audit the reports of expenditures to establish the correctness of the total amount stated, and certify as to the correctness thereof to the Navy Comptroller, Jenkins Bros., 80 White Street, New York, N.Y., for reimbursement to the New Jersey Corporation.

(d) The Navy Comptroller, Jenkins Bros., 80 White Street, New York, shall make reimbursements to the New Jersey corporation for all cash disbursements attributable and properly chargeable to the period of Navy management which are so certified by cognizant representatives of the Officer-in-Charge.

(e) All payments to the New Jersey corporation so made by the Navy Comptroller, shall be charged to Jenkins Bros., upon final settlement.

(f) All collections and other income received by branch offices, the New York Corporation, the Bridgeport plants, or the New Jersey corporation's home office which is attributable to shipments or deliveries out of inventory subsequent to 12:01 a.m. April 15, 1944, or which is otherwise properly attributable to the period of Navy management commencing at that time, will be deposited locally in accordance with prevailing instructions of Jenkins Bros. Reports of all such collections and other income shall be made periodically to the cognizant representative of the Officer-in-Charge in the same manner as reports of expenditures, and shall be accompanied by a certificate of the responsible Jenkins Bros. official to the effect that such reports include all collections and receipts attributable to the period of Navy management. Cognizant representatives of the Officer-in-Charge shall audit such reports as to the correctness of the total amount stated and shall certify the same to the Navy Comptroller, Jenkins Bros., 80 White St., New York, N.Y., as in the case of reports of expenditures.

(g) All such collections and income shall be paid over to the Navy Comptroller, Jenkins Bros., 80 White Street, New York, N.Y. Remittances of all such collections and income shall be made through orders drawn by Jenkins Bros., payable only to the order of such Navy Comptroller.

You are requested to confirm that you understand and accept these arrangements by reply to this letter.

Harold G. Bowen
Rear Admiral, USN
Officer-in-Charge

Seizure of 104 Uptown Machine Shops, San Francisco, California

Lodge 68 of the International Association of Machinists, a member of the American Federation of Labor, had been a thorn in the flesh of the people of San Francisco for a long time. It attained national prominence only, however, when it ordered that no management employing its members should work overtime unless such overtime were approved by the union. In this respect, this union introduced a new feature in labor-management relations. The fact that the management was willing to pay the customary increase in hourly wages for overtime work made no difference to the union.

The Director of Economic Stabilization, Mr. Fred Vinson (late Chief Justice of the Supreme Court) stated at a conference that this idea must be stamped out in San Francisco, and that it must not be allowed to spread over the United States. I had never before heard a civilian in government take such an unequivocal decisive stand on anything and, at the time I wondered why such a strong competent character had been held back so long by the administration. I recalled the cynical Washington expression: "Don't worry about the man in front of you. Worry about the man in back of you."

Lodge 68 had consistently refused to repeal the overtime ban, notwithstanding the orders of the War Labor Board and the urgent demands of all the Government procurement agencies and the American Federation of Labor. It was a recalcitrant union. Moreover, the situation had been greatly aggravated by the action of the business agents of Lodge 68 (Messrs. Hook and Dillon) in calling a strike at the plant of the Federal Mogul Corporation when certain members of Lodge 68 had voluntarily worked overtime at that plant in response to a work schedule which the management had posted at the invitation of the War Labor Board. The attitude of this union was strongly condemned by the American Federation of Labor and particularly so by two of its Vice Presidents, Messrs. John Frey and Harvey Brown.

I was selected by Mr. Will Davis, Chairman of the War Labor Board, to take charge of the seizure of certain machine shops in San Francisco, California, and to operate those shops overtime on war production, Lodge 68 nothwithstanding. I have always been very grateful for the support of the two well known labor leaders mentioned above, and for the fact that I was enabled to embark on this undertaking with their blessing.

Mr. Davis was most gracious—perhaps I should say forgiving, when one recalls that at Federal Ship I did not adopt Union Maintenance which was certainly the intention of the War Labor Board of which he was Chairman. He really selected me for this assignment, and when he nominated me to the Navy Department and was told that I might not be available, he told the Navy Department that he would see to it that I was made available.

In accordance with Executive Order No. 9463 and instructions from the Acting Secretary of the Navy, on the morning of 14 August 1944 I took possession of the plants of Federal Mogul Corporation, Enterprise Engine and Foundry Company, Pacific Gear and Tool Works, United States Pipe and Manufacturing Company, and Link-Belt Company (Pacific Division), all located in San Francisco, California. I was accompanied by Mr. Norwood P. Cassidy, Special Assistant to the Paymaster General; Mr. Charles C. MacLean, Jr., of the Office of the General Counsel; Captain H. S. Haislip, Assistant Chief of Staff of the 12th Naval District; Captain A. B. Court, Inspector of Naval Material for the 12th Naval District; Lieutenant (jg) S. C. Peelle, Jr., of my Washington office; and certain officers attached to the office of the Inspector of Naval Material for the 12th Naval District.

In addition to the usual copies of the Executive Order and Notice of Possession, copies of my orders requiring observance of War Labor Board Orders, prohibiting removal of books and records, and requiring overtime work, were delivered to the officials of each plant and posted at all entrances to the plants.

At the time of the seizure of the Federal Mogul Corporation plant, 111 machinists employed there had been out on strike for over two weeks. Only 14 machinists were at work, but they were complying with the overtime schedule. Accordingly, a further notice was posted at this plant, addressed to all employees of the company, which announced that possession of the plant had been taken pursuant to Executive Order of the President of the United States and that all employees were instructed to report for work immediately. A copy of this notice was mailed that day to each employee who was out on strike.

This seizure differed from all others in that management retained all its usual functions except general control and direction, particularly with respect to matters relating to labor relations. Specifically, the Government did not finance the plants nor assume any responsibility or obligations. A form of Management Agreement was devised by the Office of the Attorney General of the United States prior to our departure for the West Coast. I expected a lot of trouble with this agreement, but Messrs. Cassidy and MacLean did a wonderful and, to me, unexpected, selling job, and the agreement was soon signed by all plants.

Public relations were a most important aspect in the struggle to combat the overtime ban and to increase production.

On the afternoon of the day of the seizures and just as soon as I possibly could, I held a press conference. It did not take long to discover that all of the leading newspapers of San Francisco and Oakland were strongly behind the Government. I furnished the representatives of the press with copies of all orders and notices that had been posted at the seized plants and answered all questions relating to the seizures. At this press conference and all which followed later, I engaged in no remarks on or off the record in regard to Hook and Dillon, nor did I indulge in heroics or "wrap myself in the Flag."

At the time the work schedules were posted, I had not as yet been authorized to invoke sanctions and was therefore not in a position to advise the employees as to the various privileges that would be withdrawn from them in the event that they declined to comply with the overtime schedules. However, at 6:30 p.m. that evening (Tuesday, 15 August) I was advised by the Navy Department by telephone that the Director of Economic Stabilization, Mr. Fred Vinson, had just issued directives to the Selective Service System, the War Manpower Commission and the Office of Price Administration, directing that certain sanctions be invoked against the employees in the seized plants who were certified by the Officer-in-Charge as having refused compliance with Executive Order No. 9463, the orders of the National War Labor Board, or the orders of the Officer-in-Charge. Advance copies of these three directives were corrected in the course of this telephone conversation and, as soon as the directives had been typed, I held a late evening press conference and distributed copies thereof to members of the press. The directives were given front page publicity in the morning papers the next day (16 August) and it seems quite unlikely, therefore, that those who refused to work overtime that morning did so without knowledge of the risks involved.

The directive issued to the Selective Service System ordered it to cancel any occupational deferments and immediately induct all employees (otherwise subject to induction) who were certified by me as refusing compliance with the Executive Order, the orders of the War Labor Board, or my orders. The directive to the War Manpower Commission authorized it upon my request to withdraw from Lodge 68 all right of referral of employees in any or all of the plants and shipyards in which such lodge enjoyed the right of referral. The directive further required the War Manpower Commission to take effective action upon my request to deny further employment, including referral cards or statements of availability, to all employees who refused to work in accordance with the above-mentioned orders, and to warn all war industries not to hire such persons on penalty of having their own employees released to other industries. The directive to the Office of Price Administration directed it to utilize all lawful processes to secure a return of outstanding supplemental gasoline rations, issued on the basis of employment needs, from all employees certified by me as refusing compliance with the above-mentioned orders, and also to refuse issuance of supplemental gasoline rations to such persons.

At 9:00 a.m. on the day following the seizure, Mr. C. E.

Cox, Vice President of Enterprise Engine & Foundry Company, delivered to me a notice which Mr. Cox had seen Messrs. Harry Hook and E. F. Dillon, the business agents for Lodge 68, hand to various workmen as they left the Enterprise plant at the conclusion of the first shift on the day of the seizures. This notice bore the date 15 June 1944 when it had originally been prepared and distributed by Hook and Dillon in connection with discussions which were then current that the plants of the uptown machine shops might be seized by the Government. The notice purported to advise members of the union as to what the law required of them "in any of the phases of our present controversy between the uptown shops and Lodge No. 68." The notice referred to the provisions of Section 3 of the War Labor Disputes Act as giving the President authority to take over possession of plants in which he finds that there is an interruption in the operations which impedes the war effort, and to the provisions of Section 6 of the Act as making it unlawful for any person to induce, conspire with, or encourage any person to interfere by strike, slow down, or other means with the operation of a seized plant, or to aid any such interruption by giving direction or guidance thereto or providing funds for the conduct thereof, including payment of strike or other benefits to those participating therein. The notice then quoted and underscored the following provision of the Act: "No individual shall be deemed to have violated the provisions of this section by reason only of his having ceased work or having refused to continue to work or to accept employment."

This was the only portion of the notice that was underscored. In the notice as originally printed and distributed in June 1944, the underscored portion was immediately followed by this statement: "The portion underlined speaks for itself, and each individual member will be guided accordingly." This sentence, however, was omitted from the notice as distributed by Hook and Dillon to the employees of Enterprise Engine & Foundry Company on 14 August.

When the notice was first printed and distributed in June, it had been the subject of much discussion between the General Counsel's Office of the Navy Department and the Department of Justice. The conclusion of those participating in such discussions was that if the notice were republished following a seizure of the plants under an Executive Order, those responsible for the notice could and should be prosecuted for violation of the War Labor Disputes Act. Accordingly, as soon as Mr. C. E. Cox delivered the notice to me, I telephoned the Agent-in-Charge of the San Francisco Office of the Federal Bureau of Investigation and asked him to confer with me. Although this was the first contact between my office and the local office of the FBI, before I left Washington representatives of the General Counsel's Office had conferred with various representatives of the Department of Justice regarding the possibility that I would have need for assistance from the FBI and had been advised that the San Francisco Office of the FBI was fully acquainted with the circumstances of the case and, upon my request and following clearance with their office in Washington, would be fully prepared to place a squad of operatives at work on the case. Notwithstanding these arrangements, there was a delay of almost a day and a half before the FBI undertook to investigate the matter, and then only after repeated telephone requests by me to Washington. At 3:00 p.m. on Wednesday, 16 August, the Agent-in-Charge telephoned to state that Washington had cleared their investigation of the matter. Inasmuch as Tuesday night and Wednesday morning was the crucial time at which the employees would be determining whether they would comply with the overtime orders, and in various cases were being urged not to do so, it was apparent that during the time that the FBI was obtaining clearance from its headquarters much valuable evidence was lost. In fact, no comparable opportunity for obtaining evidence was thereafter presented. The vital importance of such evidence was later confirmed at the conference, mentioned below, between me and the Attorney General, at which the Attorney General expressed the view that the notices which had been circulated by Hook and Dillon would warrant an indictment under the War Labor

Disputes Act only when accompanied by proof of various individual acts indicating that the notices were part of a general conspiracy to induce the members of Lodge 68 to refuse to work in accordance with the overtime schedules.

The recital of events in the preceding paragraph is not made by way of complaint but rather in the hope that this experience may lead to the establishment of procedures that will prevent like delays in any future cases of a similar nature. As pointed out below, I received every material assistance as a collateral result of interviews which the FBI conducted with certain recalcitrant employees.

The overtime schedules which were posted called for the opening of the seized plants one hour earlier than required by the 8-hour schedules which had been in effect since the overtime ban was imposed on 13 April 1944. At the opening of the shops on Wednesday (16 August), which was the deadline that had been set for compliance with my orders, the overtime schedules were generally disregarded by the members of Lodge 68. The sole exceptions were the plants of United States Pipe & Manufacturing Company and the South San Francisco plant of Enterprise Engine & Foundry Company, at each of which all the employees went to work at the scheduled overtime opening hour. At the main plant of Enterprise Engine & Foundry Company only 28 out of a total of 280 members of Lodge 68 went to work at the scheduled overtime opening hour. At both Pacific Gear & Tool Works and Link-Belt Company none of the machinists worked the scheduled overtime hour. Moreover, at Federal Mogul Corporation only 29 of the 111 men who had been out on strike returned to work on Wednesday morning (16 August).

Many of the men who had reported for work late at Enterprise Engine & Foundry Company, Pacific Gear & Tool Works, and Link-Belt Company were reported to have stated that they would not work the overtime schedule unless permission to do so was granted at the regular weekly meeting of Lodge 68 which was scheduled for that evening (Wednesday, 16 August). In view of this defiance, I called a press conference and announced that I was proceeding that day to invoke sanctions against all those who had failed to comply with the overtime schedules. This announcement received wide publicity in the afternoon papers and the early evening editions of the morning papers.

Late that evening I was reliably informed that at its meeting which had just closed, Lodge 68 had adopted a resolution (a) reciting that seizure of the five shops had precluded the possibility of concluding negotiations for an agreement therewith through the medium of the Metal Trades Association; (b) resolving that Lodge 68 go on record as excluding these five shops from the application of any lodge orders or actions other than the maintenance of hours, wages, and working conditions as established in the 1942-43 agreement between such shops and Lodge 68; and (c) further resolving "that this shall be the enunciated policy of Lodge No. 68 for the guidance of its members under their individual constitutional rights employed in the shops heretofore enumerated." I was also informed that it had been determined by the union at this meeting that a special meeting of the union should be called for Sunday (20 August) for the purpose of considering the overtime ban as it affected the ninety-nine shops which had not been seized. I also heard at this time that high on the agenda for the Sunday meeting was the question whether a strike should be called in the ninety-nine shops, and that the private expectation of certain members of the union was that a strike be voted at that time.

The same evening (Wednesday, 16 August) I certified to the California Director of Selective Service the names of fifty-eight noncomplying employees in the draft ages and requested their immediate reclassification and induction. The Selective Service System moved with commendable speed and within the next few days five of the persons certified had been ordered for immediate induction and fifty others had been reclassified 1-A, of which number twenty-eight were found fit for general service and five for limited service. (After complete compliance with the overtime schedules had been obtained in all plants seized under Executive Orders

9463 and 9466, I requested the State Director to grant a stay of induction to those employees who had received notices of induction, and I took similar action in the other cases.) It goes without saying that the men in the draft ages were among the first to comply with the overtime orders. Moreover, it was reliably reported to me that the men in the draft ages had attempted actively to persuade the older men to work the overtime schedules. This action on the part of the younger men may be explained by the fact that I had indicated that I had no intention of remitting any of the sanctions invoked except possibly in the event that complete compliance was obtained in all of the seized plants.

On the second day the overtime schedules were in effect (Thursday, 17 August) compliance therewith was somewhat improved but was still highly unsatisfactory. Approximately thirty men were still out on strike at Federal Mogul Corporation, and among those who had returned there were several who failed to work the overtime schedule despite the fact that they had arrived at the plant in ample time to do so. At the main plant of Enterprise Engine & Foundry Company only forty of the 280 men on the day shift appeared at the scheduled overtime opening hour, although on the previous evening all except two of the night shift had worked the overtime schedule. At Pacific Gear & Tool Works approximately one-third of the day shift failed to report at the scheduled opening hour. At Link-Belt Company approximately 60 per cent of those who appeared for work were one hour late on the basis of the overtime schedule, and approximately 15 per cent of the force was absent. Only the plants of United States Pipe & Manufacturing Company and the South San Francisco Plant of Enterprise Engine & Foundry Company gave full compliance with the overtime schedules, as they had done the previous day.

In view of the defiance which the union had manifested at its meeting the previous evening and the continued lack of compliance with my orders, I felt obliged to take several drastic measures.

First, I sought to have the War Labor Board withdraw

from Lodge 68 all of its rights and benefits under the expired collective bargaining agreement, which rights and benefits had been kept alive only by virtue of the orders of the War Labor Board. I particularly requested (a) withdrawal of the right of preferential hiring, in order that the seized plants might be able to obtain from all available sources the workers necessary to regain lost production; (b) authority to suspend vacation schedules, in order both to restore production and meet a union threat that vacations would be taken at one and the same time, thus bringing the operations of the plants to a standstill; and (c) suspension of the particular procedure for settling grievances provided in the union agreement, in order that the operations of the plants might not be impeded by a large number of trumped-up grievances. As the result of strenuous efforts, the National War Labor Board ultimately granted the relief requested in respect of these three matters, both as to the five plants originally seized and also the ninety-nine plants seized later under Executive Order No. 9466.

Secondly, in accordance with the directive issued by the Director of Economic Stabilization on 15 August to the War Manpower Commission, I requested the Acting Director of the War Manpower Commission for the Northern California Area to withdraw from Lodge 68 all right of referral of employees in all the plants and shipyards in which such lodge enjoyed the right of referral at that time. The Acting Director of the War Manpower Commission immediately took the action requested.

Thirdly, I certified to the Regional Office of the Office of Price Administration the names of sixty noncomplying employees as a first installment, and requested the Office of Price Administration to recall any supplemental gasoline rations issued to such persons on the basis of employment needs and to refuse to issue any additional supplemental gasoline rations to such persons. (The OPA moved promptly to ascertain what supplemental gasoline rations these sixty persons held but before any effective action could be taken to recall such rations, the Administrator of the Office of

Price Administration had promulgated a regulation covering the matter, but which limited the authority of the Regional Board to persons who were certified by me as *then* refusing to comply with the Executive Order, the orders of the War Labor Board, or my orders. After receipt of this regulation in California, no action could be taken by the OPA Regional Office because all of the employees of the five companies whose plants had been seized, including the sixty employees certified on 17 August, were then complying.)

Fourthly, I requested the Selective Service System to process immediately the names certified to it the previous evening.

On the same day (Thursday, 17 August) I sought and obtained an interview with the Honorable Francis Biddle, Attorney General of the United States. We were at a critical stage in our struggle with Lodge 68. I had to make up my mind shortly about the wisdom of seizing the remaining ninety-nine uptown machine shops, because if seized it had to be an accomplished fact and a matter of public knowledge before the union meeting on Sunday morning. I was determined to tell my story to Mr. Biddle and pull no punches. Messrs. Cassidy and MacLean accompanied me on my visit. We were received in a suite in a San Francisco Hotel where the Attorney General introduced us to the Assistant Attorney General, J. P. McGranery; the United States Attorney for the Northern District of California, Mr. Frank J. Hennessy; and the Agent-in-Charge for the Federal Bureau of Investigation in the San Francisco Area, Mr. Mat Pieper.

I reported to Attorney General Biddle my progress to date with Lodge 68 and I complained that Mr. Will Davis, Chairman of the War Labor Board, had especially picked me for this job and now when I needed his support it appeared he had gone on leave with me holding the bag. Mr. Biddle was most sympathetic and immediately endeavored to reach certain parties in Washington by long distance but without success. The Attorney General confirmed our belief, as expressed previously, that some circumstantial evidence of an active conspiracy to induce the union members not to work in accordance with the overtime orders would have to be produced in addition to the notices that had been distributed by Hook and Dillon, in order to justify criminal proceedings under the War Labor Disputes Act. The Attorney General suggested to Mr. Hennessy that in the event that the FBI was able to produce such circumstantial evidence it would be well to proceed by way of grand jury rather than information. (The FBI was never able to produce such circumstantial evidence.) Mr. Pieper of the FBI indicated to the Attorney General that the FBI had no special access to information on the affairs of the union, but that they were arranging to examine the books of the union. (Such an examination was later held but no evidence of a violation of the War Labor Disputes Act developed.) Mr. Biddle assured me that the entire organization of the Department of Justice would work with me to the end that the Executive Order should be carried out and respected. He indicated that he would not be opposed to the seizure of the other ninety-nine shops. He directed the FBI to give me every possible assistance. At my request he authorized me to state to the press that he fully approved of my operations in San Francisco. This statement received prominent notice. The same editions of the newspapers carried statements by the Agent-in-Charge of the FBI that his organization had commenced an investigation of the entire matter. These announcements coming at a tense and critical time unquestionably had great influence in the successful outcome of the struggle.

The following afternoon Mr. McGranery telephoned me from Los Angeles to say that the Attorney General had been in communication with the War Labor Board and that that agency would withdraw from Lodge 68 the rights and benefits accruing to it under its expired agreement with the Metal Trades Association. Needless to say, I was delighted and deeply appreciative of the help from the Attorney General. His open, public, and forceful support of me at such a psychological moment was vital, and I doubt if I could have succeeded without it.

At 8:30 p.m. on the same day, my staff and I had conferences with the Chairman and some members of the 10th Regional War Labor Board, as well as officials of the California Metal Trades Association in regard to the desirability of seizing the ninety-nine additional shops. There was no enthusiasm for the additional take-over. Also considered at this conference were possible means of attracting additional public support for the Executive Orders and the orders of the War Labor Board. The conference closed without my indicating what further steps I intended to take in the matter of seizure.

However, I decided to beat the union to the draw. I felt strongly that the purpose of the Sunday meeting of the union was to strike the ninety-nine shops. If a strike were called prior to seizure, the difficulties of obtaining compliance obviously would be multiplied many times over. Accordingly, that evening (Thursday, 17 August) at 11:00 p.m. I sent a dispatch to the Secretary of the Navy that I should be immediately provided with another Executive Order directing me to seize the ninety-nine shops and that it was of the utmost importance that it should be in my hands as early Saturday (19 August) as possible.

The next morning, Friday, there was practically 100% compliance with the overtime orders in the five plants already seized and approximately 80% of the strikers at Federal Mogul were back at work.

Shortly after noon, Saturday (19 August), I was informed by telephone that an Executive Order providing for the seizure of the other ninety-nine members of the Metal Trades Association had been approved by all of the required agencies and was then at the White House for the President's signature, together with an order of the War Labor Board withdrawing from Lodge 68 various rights and privileges under the expired collective bargaining agreement. The text of both orders was dictated over the telephone to be released when word was received that the President had signed the Executive Order and approved the order of the War Labor Board. Great credit is due to all personnel concerned for the rapidity with which the Executive Order was submitted to the President. At approximately 2:25 p.m. the same day I was advised by telephone that the President had signed the Executive Order and approved the order of the War Labor Board.

At 2:30 p.m. I sent an identical telegram to each of the ninety-nine companies named in Executive Order No. 9466. The telegram notified the addressee that possession of its plants and facilities in San Francisco was thereby taken, effective immediately pursuant to Executive Order; directed the addressee to continue its normal business and operations subject to such orders as I should from time to time issue authorizing the posting of overtime work schedules where necessary to perform war contracts; directed all employees who were subject to such hours to work in accordance therewith commencing Tuesday, 22 August; requested a duly authorized official to attend a meeting called for 10:00 a.m. Monday, 21 August; and directed that copies of such telegrams should be posted in appropriate places in the plants.

At 3:00 p.m. I held a press conference at which I announced that the President had signed an Executive Order directing the seizure of the ninety-nine additional uptown shops by the Secretary of the Navy, and that the Acting Secretary had instructed me to take possession of such plants. Copies of the Executive Order and of the order of the War Labor Board approved by the President on 19 August 1944 were distributed to the press, together with copies of the telegram which had been sent to each of the ninety-nine companies. I stated that this action was the Government's answer to the refusal of Lodge 68 to lift the overtime ban. I also announced that the deadline for compliance with the overtime schedules would be Tuesday, 22 August, and that sanctions would be invoked against all workers who refused to obey the posted overtime hours. It was also made clear to the press that the Navy Department proposed to utilize the existing management in each shop to the fullest extent practicable. All of this received wide publicity in all editions of the Sunday morning papers. As a matter of fact, it made some editions of the Saturday afternoon papers. I had to have this publicity in time to "kill" the Sunday morning meeting of the union. Hence my unseemly haste.

On Sunday morning, between 9:00 and 10:00 a.m. and well before the time scheduled for the union meeting, acting on the advice of counsel, I took physical possession of seven of the ninety-nine shops named in Executive Order No. 9466 by posting at the main entrances of such shops printed copies of all pertinent documents. Obviously, it was impossible to seize physically all 99 shops.

The Sunday meeting of Lodge 68 was held as advertised. Its business agents issued the following statement to the press: "A motion to rescind the overtime ban was defeated overwhelmingly by a voice vote which appeared to be at least ten to one. However, all members have been informed by the meeting that no man or woman will be penalized by the Lodge for working overtime in any shops, government controlled or otherwise, if they are requested to do so and desire to comply." (The accuracy of the vote as reported has been challenged by some.) Obviously the union was not winning its fight against the United States Government.

The meeting which was called for Monday, 21 August, at 10:00 a.m. of the representatives of the ninety-nine companies named in Executive Order No. 9466, was attended by representatives of eighty-eight of such companies and also by representatives of the five companies seized under Executive Order No. 9463 and of the California Metal Trades Association. The absence of eleven of the companies named in Executive Order No. 9466 was attributable for the most part to the fact that the telegrams notifying them of the seizure and of the meeting were not delivered until Monday morning, too late for them to attend the meeting.

At the opening of the meeting an envelope containing copies of notices and orders was distributed to an official of each plant represented at the meeting. Following distribution of this material, I opened the meeting by reading a prepared statement which first outlined the reasons for the seizure of the ninety-nine additional plants and then expanded upon various points covered by a memorandum of instructions with a view to enabling the officials present to grasp at once the significance of the seizure, the purposes and advantages of the management agreement, the circumstances under which overtime might be scheduled, and the manner in which daily reports of compliance or noncompliance were to be submitted.

The management agreement was accepted almost immediately by the great majority of the ninety-nine companies and on 29 August it had been accepted by ninety-eight of the ninety-nine companies whose plants were seized on 19 August. The ninety-ninth company should never have been included—it was in a different category—and the Executive Order was subsequently modified to exclude it.

As stated above, Tuesday, 22 August 1944, was the date fixed by me for compliance with the overtime schedules in the shops seized under Executive Order No. 9466. Reports on compliance were received the following morning. They showed that on Tuesday there was full compliance in all of the shops in which overtime work had been scheduled except six. In these six shops there were approximately one hundred and twenty noncomplying employees.

Approximately half of the noncomplying employees were employed at the Pelton Water Wheel Company. On the third morning that the overtime schedules were in effect (Thursday, 24 August), and thereafter, there was full compliance by all except one of the employees of Pelton Water Wheel Company, and that one chose to quit rather than comply. He was subsequently discharged and certified by me to the War Manpower Commission for denial of referral and statements of availability and for warning to all war industries not to employ him on penalty of having their employees released. On 25 September he was permitted to return to his work upon giving satisfactory assurances of his willingness to comply.

The two other shops which had a substantial amount of noncompliance on Tuesday (22 August) were the California Press Manufacturing Company, with sixteen noncomplying employees, and the Price Pump Division of Fairbanks, Morse & Company, with twenty-nine noncomplying employees.

By Friday, 25 August, the number of noncomplying employees at California Press Manufacturing Company had been reduced to twelve. At that time I dispatched a telegram to the management directing it to ascertain immediately from each of the twelve employees and report to me whether there was any valid excuse for their noncompliance or whether they should be discharged and certified to the War Manpower Commission for denial of employment privileges. On being shown this telegram all except one made promises of compliance. The one who declined to give such a promise was discharged the next day and was subsequently certified to the War Manpower Commission. On 11 September 1944 he was permitted to return to work upon giving satisfactory assurances of his willingness to comply.

In the case of the Price Pump Division of Fairbanks, Morse & Company, the FBI instituted an investigation upon request. Thereafter a secret ballot was taken among the employees on Friday, 25 August, and the vote was twenty-four to five in favor of working the overtime schedules. Full compliance was immediately obtained.

The few other situations in which there was noncompliance, involving from one to four men, responded almost immediately to warnings of discharge and certification to the War Manpower Commission.

In addition to the two employees whose discharges are mentioned above, I ordered the discharge of seven other employees of the 104 plants seized under the Executive Orders. Five of the seven were employees of Federal Mogul Corporation who failed without proper excuse to return to work in accordance with orders. The remaining two were discharged because of willful and unjustifiable interference with the operations of the seized plants by urging other employees not to perform the duties assigned to them.

Because of the prompt and satisfactory response to the overtime schedules in the ninety-nine plants seized under Executive Order No. 9466, it was not necessary to invoke the Selective Service and rationing sanctions in the case of these plants. The nine employees discharged were not subject to the draft and held no supplemental gasoline rations.

On Monday, 28 August, I reported to the Secretary of the Navy by dispatch stating that full compliance with the overtime orders had been attained in the seized plants. The report stated that I regarded the attitude of the union officials as thoroughly unsatisfactory. Mention was made of the fact that, following the meeting of the union on Sunday, 20 August, the union officials had written each of the seized plants inviting the managements thereof to enter into separate agreements with the union and that as the result of this invitation I instructed each management not to make any agreement or commitment to the union without my prior written approval. I also reaffirmed the view, which I had previously expressed in dispatches to the Secretary that the union had in no sense repealed the overtime ban and that if possession under the Executive Orders were terminated the union would immediately invoke penalties for any violation of its policy against overtime. I stated that public reaction in San Francisco to the seizures had been highly favorable and adverted to the fact that unauthorized statements in the press on the previous day that the Navy Department was about to relinquish the plants had created consternation on the part of the plant operators. The situation was summarized by my statement that "while interruptions to production actually existing when Executive Orders were made have been removed, threat of like interruptions will continue to exist so long as overtime ban is not rescinded." The recommendation was made that consideration be given to an announcement by the Navy Department that possession of the plants would continue until the ban was expressly rescinded by the union. I also recommended that Commander H. K. Clark, U.S.N.R., attached to the Office of the Pacific Coordinator of Naval Logistics, should be designated to act for me in my absence. At that time I was also operating the seized plant of the Los Angeles Shipbuilding and Drydock Company.

On 30 August I received the following telegrammed instructions from the Under Secretary of the Navy:

IN CONFORMITY WITH EXECUTIVE ORDERS 9463 AND 9466 NAVY DEPARTMENT POSSESSION WILL NOT BE TERMINATED UNTIL LODGE 68 TAKES SUCH AFFIRMATIVE ACTION AS TO ASSURE OFFICER IN CHARGE THAT PRODUCTIVE EFFICIENCY OF THE PLANTS PREVAILING PRIOR TO IMPOSITION OF OVERTIME BAN HAS BEEN RESTORED AND WILL CONTINUE TO PREVAIL AFTER SUCH POSSESSION HAS BEEN TERMINATED X THIS REQUIREMENT WILL NECESSITATE FULL COMPLIANCE WITH ALL OUTSTANDING WAR LABOR BOARD ORDERS AND ASSURANCE THAT THERE WILL BE NO RECURRENCE OF THE CONDITIONS WHICH REQUIRED SUCH ACTION BY THE WAR LABOR BOARD AND SEIZURE BY THE PRESIDENT.

At the same time, I received by telephone from the Navy Department a summary of the four points of future action in the matter which had been agreed upon in conference between representatives of the Navy Department and the War Labor Board. Thereupon, I called a press conference and released copies of the above quoted instructions from the Under Secretary of the Navy and, in conformity with the advice received from the Navy Department, informed the press that I construed the dispatch from the Under Secretary as meaning: *first*, that the overtime ban on the seized shops must be withdrawn by formal action of the union; *secondly*, that the union must guarantee that the overtime ban would not be reinstated directly or indirectly for the duration; *thirdly*, that the union must guarantee that penalties would not be inflicted on anyone who had returned to work at Federal Mogul Corporation or who had worked overtime schedules in any of the seized shops; and, *fourthly*, that only after these steps had been taken might any collective bargaining agreement be made by the union in respect to the seized plants.

The dispatch from the Under Secretary and the interpretation placed thereon by me received considerable publicity in the afternoon papers of that day, which were distributed well before the regular weekly meeting of the union scheduled for that evening (Wednesday, 30 August). However, neither at that meeting nor at either of the weekly meetings held afterwards did the union take any action in the matter.

I was not satisfied to permit the situation to remain stabilized on the basis of Navy Department possession of the plants for the duration of hostilities if any different result could possibly be obtained. The only further step of any importance which could be taken to bring further pressure upon Lodge 68 to repeal the overtime ban was to enforce an open shop in the seized plants. I had had this step under consideration for some time. An open shop in the seized plants had in effect been authorized by the order of the National War Labor Board, approved by the President on 19 August, and the plant managements had been permitted (but not required) thereunder to operate on an open shop basis in order that they might obtain necessary workers from all available sources. I had been reluctant, however, to require an open shop because of the known aversion of the individual members of the union to working with nonunion employees. I ultimately concluded, however, that these considerations were outweighed by the possibility that the enforcement of an open shop would result in a loss of members, dues, and clearance charges to Lodge 68 on a scale sufficient to persuade it that repeal of the overtime ban would be good business.

Accordingly, on 13 September I issued from Washington Order Number 6 conveying directions to the managements of all of the plants in the possession of the Navy Department under Executive Orders Nos. 9463 and 9466 that no member, employee, or applicant for a job in any of such plants should be required to join, or remain a member of, or secure a clearance card from, Lodge 68, either as a condition of hiring or as a condition of continued employment; further, that no employee or applicant for a job should be advised either to join or not to join Lodge 68, that no management should call to the attention of any employee that he was delinquent in the payment of dues or other charges to Lodge 68 or render other assistance to such lodge in the collection of dues or charges and, finally, that no restrictions should be placed against employees or applicants for jobs in the classification of machinists voluntarily becoming members of any union, lodge, or local which was willing to admit them to membership. The order directed that copies thereof should be posted in conspicuous places in each of the plants and should be furnished to all applicants for jobs in the machinist classification.

It will be noted that Order No. 6 providing for an open shop related only to work classifications under the jurisdiction of Lodge 68 and not to those under the jurisdiction of other unions. A survey made in the middle of September 1944 indicated that there were 18 other unions that had contracts with various machine shops among the 104 shops and that these 18 unions represented over 3,100 employees in the seized shops, a number greater than the number of Lodge 68 members employed in the shops. For the most part, the other unions had agreements providing, in effect, for a closed shop as to the positions under their jurisdiction. I respected such provisions and required their enforcement in each instance where the question was raised. Relations with all such other unions were harmonious.

While I did not hesitate to deal firmly with willful infractions of my orders, the administration of labor relations in the seized shops was conducted on the basis of moderation, and consideration was given wherever individual or special circumstances warranted. Thus, all employees furnishing bona fide excuses for failure to work overtime, such as old age, poor health, or difficult transportation, were required to work only such hours as appeared proper, the determination being made in most cases on the basis of the hours worked by the individual concerned prior to the overtime ban. So also releases were granted or denied only in accordance with standard War Manpower Commission practices and procedures, except in the case of the nine individuals mentioned above who were certified to the War Manpower Commission because of their refusal to comply with the Executive Orders, the orders of the War Labor Board, and the orders from me. My office was at all times open to any employee of the seized shops desiring an interview.

At this point I should call attention again to the splendid cooperation of all concerned with the procuring of the second executive order. In view of all the departments and agencies which have to be consulted before an Executive Order can reach the White House and the length of time it usually requires to get a President's signature, the speed with which the second order was obtained was truly phenomenal. In this respect, the efforts of Rear Admiral Frederick G. Crisp and Lieutenant Commander Robert E. Bassett were invaluable. This Executive Order licked Hook and Dillon, but they were still to struggle.

There was one humorous incident which occurred while the Executive Order was being telephoned from Washington. A female telephone supervisor kept breaking in and telling me I couldn't have the line but five minutes. Finally I replied to her with considerable asperity and, I'm afraid, profanity—which fact shocked some of the listeners in the Navy Department. When asked what my priority was, I told her in good faith that it was No. 1 because I felt that there couldn't be anything more important in the world than getting that Executive Order in time to beat the union. Later on I found out that priority No. 1 was reserved for invasions!

I am indebted to the press of San Francisco for its magnificent support and also to the local representatives of the wire services as well as the New York papers. A great deal of the success realized up to this stage of the struggle was due in large part to the full and fair treatment which the Government's position received both in the news columns and the editorials of the local papers. One curious incident must be recorded.

A few days after the seizure of the five plants, Westbrook Pegler broke out in a syndicated article denouncing the seizure, beginning with my name and ending with Hitler's. This did not set well with a San Francisco which had been plagued for a decade by Lodge 68. Almost immediately the same newspaper which carried Pegler's article published a signed editorial by the publisher of a chain to which this newspaper belonged, announcing the termination of their contract with Pegler.

Of all people, why should Pegler denounce the attempt of the Government and the American Federation of Labor to break the labor racketeers? In the Navy, we would call it poor staff work. I never heard an authentic explanation of this bizarre incident, but I do know that the newspaper men of San Francisco believed that Pegler's unfortunate, ill-timed, and uninformed article was the straw that broke the camel's back.

Not being impressed by the local Navy Public and Press Relations Office, I bypassed them early in the game and used the newspaper men of San Francisco. I never met a finer outfit and frequently asked them for advice. I always gave them the whole story, and they never once violated my confidence in them. Among those to whom I owe a debt of gratitude are:

NEWSPAPER	REPORTER
Examiner	Messrs. Baum, English, Lenn
Chronicle	Messrs. Burgess, Flynn, Duerr, Griffin, Walker
San Francisco News	Messrs. Rapley, O'Reilley, Kresge
New York Times	Mr. Davies
New York Herald Tribune	Mr. Reck
Call Bulletin	Messrs. Curtin, Liddle
Post Enquirer	Messrs. Houtchins, Hardwick
Wall Street Journal	Mr. Naussauer
Christian Science Monitor	Mr. Brink
NEWS SERVICE	
International	Mr. Dunlap
United Press	Messrs. Roger Johnson, Best, Gilfoil, Hanson
Associated Press	Messrs. Schedler, Raiser

Splendid cooperation was displayed by the Honorable Roger Lapham, Mayor of San Francisco, who not only urged the repeal of the overtime ban but publicly denounced its sponsors as disloyal to the men on the fighting fronts.

Litigation Instituted by Lodge 68. During the month of September, 1944, the Lodge 68 attorneys launched a triple legal attack upon the operation of the uptown machine shops by the Navy Department, in the form of one appeal from an administrative determination, and two suits for injunctions in the Federal District Court. For approximately one month following the initial seizure, the Lodge had taken no formal action of any kind in reference to Navy Department possession, despite the fact that I had made full use of sanctions by certification of some fifty-eight employees to Selective Service, and certification of some seven employees for withdrawal of referral privileges by the War Manpower Commission. A review of my actions immediately leading up to litigation by Lodge 68 will therefore be of interest.

Review of Events Leading Up to Litigation. On 5 September 1944, I released Information Bulletin No. 1, both to the press and to the managements of the shops. This bulletin set forth the dispatch of the Under Secretary of the Navy in which it was stated that Navy Department possession would not be terminated until such time as Lodge 68 took effective action to guarantee that there would be no recurrence of the events leading up to the seizure. In a statement to the press, the dispatch of the Under Secretary was interpreted by me as laying down four conditions precedent to termination of possession, namely: (1) rescission of the overtime ban; (2) guarantee that the overtime ban would not be reimposed; (3) guarantee of no retaliation against union members working overtime or refusing to strike; and (4) consummation of satisfactory agreements with the uptown machine shops.

Thereafter, on 8 September 1944, the Acting Officer-in-Charge ordered the discharge of one Arthur B. Burke, by his employer, the National Motor Bearing Company, Inc., and certified the name of Burke to the War Manpower Commission for withdrawal of referral privileges. Burke was a member of the Lodge 68 shop committee at the company and, while not shop steward, was the most active member of the committee from the union standpoint. He was the first employee holding official status with Lodge 68 to be discharged and certified. On 9 September 1944, the Acting Officer-in-Charge released Information Bulletin No. 2, which stressed the necessity of obtaining Navy approval for release of members of Lodge 68; warned against interference with other employees on jurisdictional grounds; and indicated that the grievance machinery provided by the Lodge 68 contract would thereafter be disregarded except at the first level.

On 12 September 1944, the Acting Officer-in-Charge ordered the discharge of Martin A. Joos by his employer, Bodinson Manufacturing Company, and certified his name to the War Manpower Commission for withdrawal of referral privileges. Joos was the shop steward of Lodge 68 at Bodinson Company, and was a long-time member of the Lodge with pronounced union views. On 13 September 1944, I transmitted Order No. 6 by dispatch from my Washington Office, which was immediately released to the press by the Acting Officer-in-Charge. This order, while technically only supplementing the order of the National War Labor Board of 19 August 1944, nevertheless indicated forcibly for the first time the determination of the Navy Department to operate the machine shops of the seized plants on an "open shop" basis. As a result of the foregoing, the business agents of Lodge 68 were faced both with loss of their control over the personnel of the shops through abrogation of their preferential hiring privileges, and with denial of their participation in grievance procedure. They likewise undoubtedly felt under pressure by their membership to prove that they could still provide protection for their shop stewards or shop committeemen who, under ordinary circumstances, would have been practically free of any threat of discharge except with union consent.

Summary of Lodge 68 Actions. The first step by Lodge 68 was taken 13 September 1944 in the filing of an appeal to the War Manpower Commission in behalf of Arthur B. Burke from denial of referral by the United States Employment Service. This appeal was noted in accordance with usual War Manpower Commission's appeal procedures and sought to contest the existence of any valid ground for Burke's discharge. The next step was taken by Lodge 68 on 20 September 1944, in the filing of a suit in the Federal District Court in the name of Martin A. Joos against a large number of defendants, including the Secretary of the Navy, the Chairman of the War Labor Board, the Chairman of the War Manpower Commission, the local agents of certain of the foregoing, me, my Acting Officer-in-Charge, and the Bodinson Manufacturing Company. The suit sought to enjoin the defendants from withholding a release from Joos and further asked a mandatory injunction to compel his referral to other work in war industry. The third action by Lodge 68 was taken on 25 September 1944, in the filing of a suit against substantially the same defendants as in the Joos suit, but this time in the name of the President and Secretary of the Lodge in behalf of all its members. This suit asked an injunction against (1) discharging any members of the Lodge, (2) denying right of referral to any discharged member of the Lodge, and (3) putting into effect the "open shop" policy by the enforcement of Order No. 6 of the National War Labor Board Order of 19 August 1944. The suit likewise asked a declaratory judgment that the Labor Board Order of 19 August 1944 was invalid (1) because it was issued without notice or hearing to Lodge 68 and (2) because it sought to deprive the Lodge of rights which were not an issue in dispute between the Lodge and the uptown machine shops.

Appeal of Burke. Upon the noting of the appeal by Burke, local officials of the War Manpower Commission advised the Acting Officer-in-Charge that, in their opinion, normal appeal procedures were available in the case, but that no action would be taken until determination by the Chairman of the War Manpower Commission whether he wished to assume personal jurisdiction of the case on the first appeal level, as he was authorized to do by existing regulations, if they were applicable. Shortly thereafter, the Chairman determined to assume jurisdiction of the appeal, but announced that evidence would be taken before a special panel to be convened in San Francisco, which would forward its findings and recommendations to Washington for the final action. Inasmuch as there was no provision of law by which the Department of Justice or the United States Attorney's Office could appear in behalf of the Navy Department at such an administrative hearing, it was deemed advisable for the Navy Department to take active charge of the defense of the appeal, and arrangements were made for the return to San Francisco of Charles C. MacLean of the Office of General Counsel.

Suit of Joos. Upon the filing of the suit by Martin A. Joos on 20 September 1944, immediate service was effected on Commander H. K. Clark, USNR, the Acting Officer-in-Charge. No service was effected upon me, as I was then in Washington. Concurrently with the filing of the suit, plaintiff's attorneys asked a temporary restraining order, which, under the guise of maintaining the status quo, actually would have had the effect of granting the plaintiff all the relief to which he would have been entitled on a final hearing. The restraining order, which was granted by Judge Michael Joseph Roche immediately upon the filing of the suit, without notice to any defendants, directed the Navy Department and Bodinson Manufacturing Company to grant a release to Joos, and directed other defendants to issue him a referral for work in war industry. Service of the restraining order was effected upon Commander Clark, together with the notice of the suit.

It was the view of the Acting Officer-in-Charge that he was subject only to the orders of his superior officer, over whom no jurisdiction had been obtained, and that the restraining order should not be complied with in the absence of direction to such effect from his superior officer, since it would have granted Joos all the relief pleaded for without a hearing to the defendants, and would have disturbed the status quo, making it thereafter impossible to reimpose sanctions against Joos.

On the morning of 21 September 1944, advice from Mr. H. Struve Hensel, General Counsel of the Navy Department, was to the effect that the restraining order had been issued without justification, and that the terms thereof should not be complied with, either by the Acting Officer-in-Charge or by Bodinson Manufacturing Company. Later in the day, the United States Attorney conferred with Judge Roche, pointing out that the restraining order had been issued without the posting of the required statutory bond and, presumably, also pointing out the serious implications of the order, affecting as it did the official conduct of high-ranking officers of the Government in Washington. An informal understanding was then reached between Judge Roche and the United States Attorney to the effect that the restraining order would not be enforced, and should be considered merely as a rule to show cause. The hearing on the issuance of the temporary injunction was likewise continued until Monday, 2 October.

Policy Determination in Washington. The noting of the Burke appeal and the filing of the Joos suit raised questions of high governmental policy, not only affecting the Navy Department operations of the uptown machine shops, but also raising serious problems of the entire governmental program of wartime manpower controls. Accordingly, over the period from 21 September to 23 September 1944, a series of conferences were held in Washington by officials of the Department of Justice, the War Manpower Commission, the War Labor Board, the Office of Economic Stabilization, and the Navy Department. The chief ground of contention lay in the interpretation of the directives of the Director of Economic Stabilization, Judge Vinson, to the Chairman of the War Manpower Commission, The Honorable Paul McNutt, dated 15 August and 21 August 1944. The first question was whether action taken under the directives was subject to appeal under normal War Manpower Commission appeal procedures. The second question was whether the directives added anything of weight to the normal regulatory powers of the War Manpower Commission or whether they were merely declaratory of the fact that such normal procedures could be availed of in the case of the seized shops.

As an immediate result of the conference, I agreed, upon the advice of the Office of the General Counsel, to interpret the 15 August and 21 August directives as not depriving members of Lodge 68 of their right to appeal from denial of referral in accordance with normal War Manpower Commission regulations. In order to formalize this agreement, I transmitted a letter to the Chairman of the War Manpower Commission under date of 23 September.

Following the filing of the second suit by Lodge 68 on 25 September, conferences were resumed on the general subject of interpretation of the War Manpower Commission sanctions. The Navy Department, the War Manpower Commission, and the Director of Economic Stabilization took the position throughout that the directive of 15 August and 21 August went beyond normal War Manpower Commission procedures in three respects, namely: (a) that the directives authorized denial of referral following a discharge for noncompliance with the Executive Orders, the orders of the War Labor Board, or my orders, whereas normal procedures allowed a referral to any discharged employee (with the exception of the practically never used theory of discharge by "constructive quit" or for the effective prosecution of the war); (b) that the directives authorized denial of referral for an indefinite period which might be as long as the duration of the war, whereas normal regulations restricted denial of referral to a maximum of sixty days; and (c) that the use of the directives in any case offered to the War Manpower Commission the option of determining whether to proceed under the directives or to apply normal regulations. The position of the Navy Department was opposed by the Justice Department and the War Labor Board because of the policy of both of these agencies to avoid any legal defense which necessitated upholding the legality of action by Executive Order. I never understood why the Justice Department was so timid about testing in court an Executive Order issued in time of war. Consistent with this attitude, the Justice Department in particular sought to secure a united Government position that the directives were of no substantial effect and did nothing more than advise the Navy that it could utilize normal War Manpower Commission procedures. The final result of the conference was a compromise agreed to by all parties, by which the denial of referral pursuant to the directives was limited to a period of sixty days, but under which it was definitely conceded that the directives imposed sanctions going beyond normal War Manpower Commission procedures. The compromise reached at the conference was embodied in an exchange of letters between the Chairman of the War Manpower Commission and the Director of Economic Stabilization, both dated 29 September 1944. It is of particular importance to note that, while the letter from the Chairman of the War Manpower Commission does not expressly so state, it was agreed that any certification by me under the directives would be accepted by the War Manpower Commission as prima facie proof of the fact of noncompliance, to be rebutted only by evidence from the discharged employee sufficient to overcome the presumption which had thus been raised against him.

Authority to Order Overtime. The certification of Joos had been primarily based upon his interference with the work of other employees. Going far beyond his permissible duties as shop steward, he had sought directly to interfere with members of other unions in the performance of work which he alleged to have been under the jurisdiction of the machinists' union. Joos had likewise, however, indicated a recalcitrant attitude towards the overtime hours scheduled at my direction. While purporting to comply with these orders by making arrangements to work a nine-hour day, he had absented himself without permission for a sufficient number of hours each week so that his total hours had been less than forty-eight during each week following the seizure. Therefore, upon the certification of Joos' name to the War Manpower Commission, the Acting Officer-in-Charge had assigned two grounds for the denial of referral, namely, the interference with other employees and the refusal to work overtime.

In preparation for the hearing of both Lodge 68 suits, Alfred S. Berg, Esq., Special Attorney of the Department of Justice, arrived in San Francisco on 30 September to aid Assistant United States Attorney William B. Licking in the defense. Upon his arrival in San Francisco, Mr. Berg immediately raised the question whether I had authority to order overtime to be worked by an individual employee, and whether, upon the refusal of any employee to work schedule overtime, this ground could be assigned by me as basis for discharge and procedure under the directives of the War Manpower Commission. Mr. Berg took the view that Executive Order 9370, pursuant to which the directives had been issued, authorized action solely as a result of failure to obey orders of the War Labor Board, and made no reference to failure to comply with the Executive Orders or my orders. He further pointed out that the War Labor Board order of 3 June 1944, the noncompliance with which had led to the seizure, had expressly recited that such order would not compel any employee to work overtime against his wishes. It was his view that certification to the War Manpower Commission could be employed solely in the event of a concerted refusal to work overtime, which would be indicative of the continuance of the overtime ban by Lodge 68, and thus of a continued noncompliance with the War Labor Board order of 3 June 1944, which directed a rescission of the overtime ban. The Acting Officer-in-Charge and Mr. MacLean immediately advised Mr. Berg that this view was unacceptable to the Navy Department, and that it would be impossible to effectuate the purposes of the Executive Order unless I had authority to enforce the working of overtime by direct sanctions upon the individual employee.

The General Counsel of the Navy Department took the position that, not only because of past conduct of this activity but even as a *de novo* matter, the Navy Department would insist that certification to the War Manpower Commission was warranted by the refusal of an individual employee to work overtime. The views of the General Counsel were communicated to Mr. Levy, First Assistant to the Assistant Attorney General in Charge of Claims, who, while voicing doubts about the position originally taken by the Navy, nevertheless agreed that matters had progressed so far that the Department of Justice would be forced to accede to this position. As a result of various conferences and of the understanding reached by governmental agencies, it was apparent that, while the Navy had been forced to recede to some extent from certain positions taken by it, nevertheless, interpretations had been arrived at which continued to furnish a basis of effective operations of the uptown machine shops under the Executive Orders.

Hearing of the Burke Appeal. The hearing of the appeal of Burke was held before the special panel of the Regional War Manpower Commission on 29 September and 4 October. Testimony was received from Burke, from the Lodge 68 shop steward and the other member of the Lodge's shop committee, and from employees of the company. The facts on which the Acting Officer-in-Charge had based the discharge and certification of Burke, as fully established at the hearing, were as follows: A young journeyman machinist in the employ of the company, one Tony Martin, had been assigned to work as a helper to Ernie Zeller, an experienced tool and die maker. Several months prior to the seizure, the union had complained that Martin was performing work which should be performed only by a tool and die maker, and not by a machinist. Since Zeller was a long-time member of the union and former official thereof, and in view of his experience in the trade, it was agreed that he alone should determine the work which Martin should perform. The agreement worked with satisfaction and without incident until shortly following the seizure, at which time Burke again made a complaint both to Martin and to Zeller of the nature of the work which Martin was performing. His complaints followed shortly after a visit to the plant by Ed Dillon, one of the union's business agents, who admitted at the hearing that this was the only plant which he had visited following seizure. Upon report of Burke's attempted interference, I issued instructions, which were communicated to Burke, that Martin should continue his assigned work without interference. Following receipt of these instructions, Burke directly interfered with Martin by ordering him to stop certain work which he was performing under instructions from Zeller. At the request of the Acting Officer-in-Charge, Federal Bureau of Investigation operatives had investigated the foregoing circumstances. The investigation bore out the truth of the accusations against Burke, and the FBI reported to the United States Attorney that Burke had been guilty of a technical violation of the War Labor Disputes Act. It was only after this report by the FBI that the Acting Officer-in-Charge had ordered the discharge and certification of Burke. While the United States Attorney had subsequently declined prosecution of the criminal charge, it was not felt that this had in any way lessened the propriety of the discharge and certification.

FBI Assistance. Burke was represented at the hearing by the attorney for Lodge 68. It was obviously the tactics of the defense to so direct the evidence that it would appear that Burke's interference with Martin had occurred only prior to and not subsequent to the receipt of knowledge by him that I had issued an order as to the performance of assigned work by Martin and the non-interference with such work. The investigation by the FBI had produced signed statements by Burke and other members of the shop committee clearly indicating that Burke's actions had occurred subsequent as well as prior to my orders. It was of particular importance that these signed statements be made available at the hearing in order to prevent a change in testimony by Burke and other employees. The local office of the FBI refused to release the signed statements, or to permit their agents to testify, without direct instructions from Washington. Mr. MacLean therefore communicated with Mr. H. Struve Hensel, who in turn communicated with Mr. Hugh Cox of the Attorney General's Office. At the instance of the latter, instructions were forwarded from Washington to make available the signed statements and also to authorize the agents to testify. The signed statements were introduced in evidence, and accomplished the necessary purpose of establishing the date of the interferences by Burke. It was not necessary to call upon any of the agents to testify. At the conclusion of the hearing, the War Manpower Commission panel took the case under advisement and the report of the panel was not received for some two weeks.

Hearing of Joos Suit. The suit of Martin A. Joos came along for hearing on the motion for temporary injunction and was argued before Judge Roche on 2 and 3 October. The chief ground of defense was that Joos had failed to avail himself of his administrative remedy by appealing to the War Manpower Commission. It was further argued that Joos had not shown any such injury as to justify temporary relief or equitable cognizance; that the suit was prematurely brought; and that there was no likelihood that plaintiff would prevail on the merits because of lack of jurisdiction over the defendants and the impropriety of controlling the discretion of executive officers by injunction. The basic argument of Joos was that the members of the War Manpower Commission had stated that they would not render any decision in any matter pertaining to Lodge 68 or the members thereof who had been discharged under my orders. No factual ground whatsoever was stated for this allegation, nor did it appear that Joos had applied for a referral to any office of the United States Employment Service. In rebuttal of these allegations, the court was furnished with my letter to the Chairman of the War Manpower Commission in which I stated my understanding that usual appeal procedures would apply in all cases of certification. The court was also advised that the Chairman of the War Manpower Commission, by telegram dated 23 September 1944, had notified Joos of his right to utilize appeal procedures. The Chairman of the War Manpower Commission furnished an affidavit setting forth the War Manpower Commission appeal regulations and pointing out the applicability of these regulations to Joos. The court was likewise furnished with affidavits of the Acting Secretary of the Navy setting forth the importance of the production of the uptown machine shops in the war effort and of the Acting Officer-in-Charge setting forth the basis upon which he had taken action against Joos. During the course of the hearing, Judge Roche indicated an opinion that the action was prematurely brought inasmuch as plaintiff had not sought to appeal from denial of his release, or to obtain referral from the United States Employment Service. Accordingly, at the conclusion of the hearing on 3 September, an understanding was reached between counsel that Joos would immediately note an appeal with the War Manpower Commission, and that the court action would be continued until such time as it has been determined whether the appeal procedure furnished plaintiff with an adequate remedy.

Withdrawal of Joos Certification. Joos immediately noted an appeal to the War Manpower Commission, which was thereupon set down for hearing on 9 October. The evidence upon which the certification of Joos had been based had been obtained by personal interview by the Acting Officer-in-Charge and a member of his staff with Messrs. Bodinson and Kelly, President and Vice President of the company, certain employees of other unions, and other officers of the company. The facts as originally developed had appeared fully to warrant the action taken. Upon investigation preparatory to the hearing, Messrs. Bodinson and Kelly were unable to substantiate fully the charges made by them during previous interviews. It appeared that they had been to some extent activated by personal prejudice against Joos, and that their desire had been to effect his discharge under the protection of the Navy Department. What had appeared to be cases of direct interference with the work of other employees developed upon further investigation to be little more than over-talkative conduct on the part of Joos which had not led to any actual work stoppages. Furthermore, while it still appeared probable that Joos had pursued a course of deliberate violation of the overtime orders, the available testimony was not such as to make a clear and convincing case against him. His conduct was not so serious as that of Burke, and did not appear to warrant denial of referral for so long a period. There was precedent in other cases for the withdrawal of certification prior to the full 60 days. Upon consideration that the denial of referral would have extended for four weeks by the date set for the hearing, the Acting Officer-in-Charge recommended that it would be appropriate to advise the War Manpower Commission that certification was terminated on that date, on the ground that a four-week period of denial of referral was appropriate in relation to violations of orders committed by Joos. This course was concurred in by Mr. MacLean and Mr. Berg, who felt it would be inadvisable to endanger the enforcement of sanctions by proceeding to hearing on a case so lacking in proof. I was advised of this recommendation and approved the form of letter to be transmitted to the War Manpower Commission. Such letter was furnished to Joos and to the War Manpower Commission at the time set for the opening of the hearing on 9 October. The Commission thereupon dismissed the appeal as moot. (In the continuance of the court hearing on 16 October, Judge Roche likewise dismissed the case as moot.)

Concurrently with the withdrawal of certification of Joos, the Acting Officer-in-Charge released to the press a statement concerning such action, stressing, however, that the action in no way implied the lack of the continuing of my power to exercise sanctions in proper cases, and further pointing out that the decision did not mean that Joos would be reemployed in any establishment in the possession of the Navy Department. While the attorney for Lodge 68 made statements to the press implying that the Navy Department had "backed down" because its prior actions had been untenable, the press accounts fairly reported the statements of the Navy Department in this respect. Immediately upon withdrawal of certification, a mimeographed letter was addressed to all the shop managements, instructing them not to rehire Joos. The management of Bodinson Manufacturing Company was specifically instructed in this respect and, upon the immediate reapplication of Joos for employment, notified him that there was no position available.

Lodge 68 Suit. The second suit, which had been filed in behalf of all of the members of Lodge 68, was continued from time to time until 16 October. Argument began on that day and was continued on 17 October with Messrs. Licking and Berg, of the Department of Justice, representing the defendants. The principal grounds interposed by the defense were (a) lack of jurisdiction over certain indispensable defendants, (b) lack of jurisdiction over the subject matter, (c) amount of controversy less than $3,000, and (d) failure of the company to state a cause of action. It was considered that the presentation of the case by Messrs. Licking and Berg was most effective, and a letter of appreciation in that respect was afterwards transmitted to them by me.

On 23 October 1944, Judge Goodman announced his decision in the Lodge 68 suit. He dismissed the complaint as to all defendants, and over the strenuous objections of plaintiff's attorney, denied leave to amend the complaint. The oral opinion of Judge Goodman, as later revised for publication held as follows:

(1) The equities claimed by Lodge 68 are most unconvincing in the light of its persistent defiance of government orders issued in aid of the war effort.

(2) No "justiciable" controversy exists, since the Lodge 68 complaint is based upon anticipation of the possible future course of action of the defendants. Of the two specific instances of discharge and withholding of referral alleged, one is now moot and the other in process of administrative determination.

(3) If the court were to rule on the validity of the War Labor Board order of 19 August 1944, it would hold

it valid, though issued without hearing, since issuance had been preceded by long and thorough investigation of shop employment conditions and in any event, a hearing was not mandatory under the statute.

(4) If the court were to rule on the issue whether denial of clearance and referral was a "penalty," it would hold that the same was not a penalty but was remedial. Relief would be granted to the plaintiff only if it appeared that a refusal of clearance in a particular case were not reasonably related to the objectives, the accomplishment of which are by law and necessity committed to the executive.

The decision of the court was immediately released to the press and resulted in favorable notices.

Decision in Burke Appeal. Meanwhile, the War Manpower Commission appeal panel had, on 13 October, completed its findings and recommendations. While the panel found the facts of the case to be in accordance with the contentions of the Navy Department, and as outlined above, it recommended that clearance and referral be granted to Burke, on the ground that his conduct did not indicate a deliberate intention to violate the War Labor Disputes Act and Executive Order, or knowledge that his conduct was in violation thereof. The panel had authority only to make a recommendation for submission to the Chairman of the War Manpower Commission. Because of the adverse recommendation of the panel, however, it was deemed advisable to submit a brief on my behalf to the Chairman. Accordingly, a brief of some twenty pages was prepared by Mr. MacLean in which the evidence was carefully reviewed and in which the importance of the decision to the Navy Department operation of the seized shops was strongly urged. It was pointed out in particular that the case was of importance from the standpoint of its effect on subsequent compliance by the employees of the seized plants.

On 1 November, the Chairman of the War Manpower Commission handed down the decision, which was immediately released to the press. The decision reversed the recommendation of the panel and fully upheld my action in the discharge and certification. The decision likewise laid down the principle that conduct in violation of the Executive Order, my orders, or the orders of the War Labor Board, should be deemed as intended to provoke a discharge, and thus as a "constructive quit." In accordance with this principle, the Chairman stated that denial of clearance could extend for the same period as that applicable in the case of "a constructive quit," namely, for sixty days. In the instant case, the Chairman held, however (in view of the fact that it was a case of first impression), that the denial of clearance in the case of Burke had already extended for a sufficient period and that, upon indication of his intention to comply with lawful orders in the future, he could be reemployed in one of the Navy Department shops, if I so elected, or if not so reemployed, could be granted referral elsewhere.

On 2 November, the Acting Officer-in-Charge issued a mimeographed letter to all shops instructing them not to reemploy Burke. The latter subsequently appeared at the plant of National Motor Bearing Company, Inc., but was informed that no position was available to him. A check of the seized shops several weeks thereafter indicated that he had not applied for reemployment in any other of the Navy operated plants. However, on 8 November, Pacific Can Company reported that they had inadvertently employed Joos, in disregard of the express instructions of the Acting Officer-in-Charge to that effect. Inasmuch as the employment of Joos had been an accomplished fact for several days prior to notice to this office, it was determined that his employment should be continued, so long as he conducted himself properly.

Summary of Legal Principles Established. It may be advisable to summarize at this point the legal positions established as a result of the two court actions, the administrative appeal, and the understandings reached by the various government agencies:

(1) The validity of the War Labor Board order of 19 August 1944, withdrawing preferential hiring and grievance procedure rights from Lodge 68 was upheld by court decision, despite the lack of hearing prior to issuance of the order.

(2) The validity of the use of sanctions under Executive Order 9370, and in particular of certifications to the War Manpower Commission for the withholding of clearance and referral, was upheld by the court.

(3) The court decision upholding the issuance of sanctions under Executive Order 9370 nevertheless indicated that the court would grant relief, if it should appear that there was an abuse of power under such sanctions, for purposes not directly related to the Executive Order.

(4) Certification by me to the War Manpower Commission was agreed as furnishing sufficient prima facie evidence to the War Manpower Commission to justify the denial of referral at the first level.

(5) The right of persons certified to the War Manpower Commission to appeal from denial of referral in accordance with normal War Manpower Commission regulations was established.

(6) The maximum period of denial of referral was fixed at sixty days.

(7) The principle was established by the War Manpower Commission that denial of referral could terminate at any time prior to the sixty days period if the recalcitrant employee furnished satisfactory evidence that he was no longer refusing to comply with lawful orders. (The nature of the evidence required to be presented by the employee in this respect was not made clear.)

(8) My authority to order overtime and to enforce the performance thereof by individual sanctions was not weakened.

(9) No successful challenge was made to the existence of the fact of constructive possession of the seized plants by the Navy Department in accordance with the procedures followed under Executive Order 9466.

It may be emphasized that all the foregoing conclusions were matters of first impression, either as to court determination or as to administrative decision.

Tightening of Control over the Seized Shops. Both during and following the pendency of the court and administrative proceedings, every effort was made to tighten up the control of the seized shops by the Navy Department, and to restrict any conduct by Lodge 68 which might hamper their operation. On 26 September, managements were notified for the first time that they might consider scheduling overtime in excess of that prevailing prior to 13 April 1944, in the event that essential war commitments appeared to require this action. On 9 October the Acting Officer-in-Charge issued Order No. 7, providing that managements could employ persons who were also employed elsewhere on a full-time or part-time basis, despite any union ban on this practice. This order was specifically directed against a ban which had been in effect by Lodge 68 ever since the institution of the "victory worker" program in 1942. Any such ban was deemed incompatible with the full utilization of manpower in the war effort.

Release of Orders 8 and 10. During the pendency of the court proceedings, officers of the Department of Justice requested that no action be taken by me which would in any way affect the *status quo*. With the successful conclusion of the court proceedings, the Acting Officer-in-Charge on 7 November made simultaneous release of Orders Nos. 8 and 10, which provided in substance as follows:

Order No. 8 directed that terms and conditions of employment in effect at the time of the seizure should continue unchanged except with respect to those rights of Lodge 68, which had been withdrawn by order of the National War Labor Board of 19 August. This order likewise established grievance procedures with respect to Lodge 68 and other unions represented in the plant. In the case of Lodge 68, it was ordered that the shop steward could continue to present grievances at the first level, but that if no agreement were reached at that level, the matter should immediately be reported to me for such action as I might deem proper. The order prohibited the business agents of Lodge 68 from direct contact with company management in grievance matters. In order, however, to avoid conflict with the National Labor Relations Act, it was provided that the business agents of the lodge, on application by them, could confer with me on grievance matters. In the case of unions other than Lodge 68, Order No. 8 recognized their contractual rights by providing that grievances could be disposed of in accordance with the procedure in effect at the date of seizure. The provisions of some of said agreements provided for arbitration at the final level.

I deemed it incompatible with governmental sovereignty to permit such arbitration and accordingly provided that in any case where the parties were unable to dispose of the grievance by mutual agreement, the matter should forthwith be reported to me "for such action as I may determine to take." The language of the order avoided any direct abrogation of arbitration procedures, since it was feared such an order might be regarded in some quarters as in conflict with the "freezing" provisions of Section 4 of the War Labor Disputes Act.

Order No. 10. This order (1) prohibited any unauthorized person from interfering with any employee in the performance of work assigned to him by his superiors, with an express warning to shop stewards or committeemen against such conduct; (2) ordered all employees to perform assigned work, regardless of any question of craft or classification; (3) called to the attention of all employees the penal provisions of Section 6(a) of the War Labor Disputes Act; and (4) warned employees that any violation of the prior provisions of the order might subject them, at my discretion, to immediate discharge and to the imposition of all authorized sanctions, including cancellation of occupational deferment, withdrawal of supplemental gasoline rations, and certification to the War Manpower Commission for refusal of clearance and referral. The necessity of this order became apparent when it was learned that the Northern California Area Director for the War Manpower Commission did not enforce the national rule in reference to the withholding of referral in the case of a "provoked discharge," but held that clearance must immediately issue in the event of such a discharge. By the promulgation of Order No. 10, I made it possible to certify to the War Manpower Commission for denial of clearance and referral those employees whose conduct in other areas of the United States would have rendered them subject to the usual penalties for provoked discharge.

A provision of Order No. 6 directed that management should not render assistance in the collection of dues. Evidence indicated that this provision had been even more effective in its operation with respect to the position of Lodge 68 than the elimination of the requirement of clearance. Prior to Order No. 6, it was only necessary for the business agents of the lodge to complain to an employer of delinquency in dues, and the offending member was ordered immediately to pay the dues in arrears or to be discharged. Many persons who did not wish to offend the lodge by deliberately refusing to take out clearance cards were reluctant to pay their dues. The union in turn, so evidence would appear to indicate, was reluctant to expel members for delinquency during the period of Navy operation. Information from shop stewards in one or two instances indicated that delinquency in dues ran as high as 75% of the union members in certain plants. Undoubtedly this had an effect upon the lodge, but it did not induce action on its part.

Business agents of the lodge lost to a certain extent the control over their membership in the seized shops. Formerly the business agents were accustomed to walk through the plants at will. They were not now permitted to do so and did not even seek admission except in two or three instances. Business agents were likewise formerly accustomed to handle all grievances for their membership, occasionally even at the first level. This they had also been prohibited from doing through direct relations with management, nor were they willing to take up grievances at the second level with me.

Their only direct participation in the affairs of the seized shops was their representation of members of the lodge in hearings before appeal committees of the War Manpower Commission on denials of release. Here again, there was no evidence that this loss of control by the business agents had been sufficient to lead to the likelihood of action on their part.

Relations with Other Unions. During the initial period of the seizure, the entire attention of the staff was devoted to relations with Lodge 68, and to some extent with the membership of Local 1330, whose members had associated themselves with Lodge 68 in connection with the overtime ban. No effort was made to police the relations of other unions with the managements of the seized shops. Three subsequent steps, however, led to greatly increased contacts with other unions: the expansion of release procedures to include the members of all unions; the greater contact with the shops resulting from the assignment of two permanent liaison officers; and the requirements of Order No. 11 (hereinafter discussed), by which no negotiations for or changes in labor union contracts could be effected without my consent. With the exception of the walkout at Schmidt Lithograph Company, hereinafter noted, relations of my staff with the officers and members of other unions were harmonious to an exceptionally high degree. The chief difficulty was one of conveying to the officials and members of all the unions the fact that they were subject to obligations under the War Labor Disputes Act and my orders to an equal degree with members of Lodge 68. When this fact was communicated to them, cooperation of a high degree was generally offered.

The great majority of the 60 or more unions represented in the seized shops had closed shop contracts or understandings. It had been the past practice of union officials to demand an immediate discharge of any employee not in good standing with their union because of delinquency in dues or otherwise, and in many instances they assumed to order the affected employee to quit his job prior to taking the matter up with the management. I took the firm position from the outset that, while the provisions of closed shop contracts would be respected, no union official would be permitted to give direct orders to employees of the seized shops and, in particular, to give any orders which would result in interference with the work of the shops. In these few instances, therefore, in which union officials, ignorant of the incidents of Navy seizure, assumed to force delinquent union members off the job, I promptly notified the union officials concerned that such conduct was in violation of the War Labor Disputes Act and would result in the imposition of drastic penalties if continued, or if not corrected. When this fact was communicated to the union officials, they cooperated with me in every instance by agreeing to desist from direct orders to their members employed in the seized shops and to take up such matters with the managements for submission to me.

Walk-out at Schmidt Lithograph Company. The one serious instance of conduct in violation of the War Labor Disputes Act and of my orders occurring on the part of a union, other than Lodge 68, was a walk-out at the plant of Schmidt Lithograph Company, which was incident to a city-wide walk-out on 21 November 1944 by members of Local 17, Amalgamated Lithographers of America. The walk-out stemmed from a decision of the War Labor Board denying a retroactive pay increase which caused much dissatisfaction among the members of the union. The walk-out began without sanction of union officials, as a spontaneous action on the part of employees of certain lithographic companies not included under the seizure. A few of these employees from other plants congregated in front of the Schmidt Company premises at the noon hour and induced the members of Local 17 employed at the Schmidt Company to join in the walk-out during the afternoon.

After the union members were apprised of the situation, the affair was amicably settled and they went back to work.

The union leaders were of great assistance in handling this case.

Following the walk-out at Schmidt Lithograph Company, the Acting Officer-in-Charge instituted steps to apprise all managements and employees more fully of their obligations under the seizure orders. To this end, all union business agents coming into contact with this office were given an explanation of the aspects of the seizure which affected their membership and were furnished with a set of the executive orders and my orders. The Acting Officer-in-Charge likewise addressed a mimeographed letter for the attention of the shop stewards and members of all unions other than Lodge 68. This letter was either posted upon the bulletin board or handed personally to shop stewards of other unions in all plants and served to create a much better understanding between the employees of the shops and the staff of my office.

As the business agents of the various unions, other than Lodge 68, became increasingly familiar with my authority and my position in relation to company management, many of them adopted a policy of taking up important matters directly with me, prior even to discussion thereof with the management of the companies. This had been especially the case with respect to certain companies in whose management personnel the business agents did not appear to have full confidence. Business agents generally appeared to understand that they would receive fair and impartial treatment from my office, and that they could rely on the carrying out of any commitments made by me on behalf of company management. Indeed, in a proposed labor union contract submitted to me for approval, the labor union had approved a provision which would have designated me as arbitrator with final decision on all grievances during the period of Navy operation. (This contract provision was not accepted, however, because of my belief that I should not formally be designated in such capacity.) These relationships with the business agents were cited because it was felt to be of great importance to successful operation of the uptown shops that the Navy Department not acquire the reputation of being anti-union or anti-labor. Of necessity, the operation entailed a very firm attitude towards Lodge 68, including the insistence on open-shop policy. Unless great care had been exercised by me and my staff, the impression might have been created that the attitude towards Lodge 68 was an attitude of opposition to other unions as well.

War Labor Disputes Act. The War Labor Disputes Act provided in Section 4 thereof that the terms and conditions of employment in effect at the time that possession of a plant was taken should continue in operation unless application for change therein was filed with the War Labor Board pursuant to Section 5 of the Act. Section 5 of the Act provided that either the government agency operating the plant, or a majority of the employees or their representatives, might file an application with the Board for a change in wages or other terms and conditions of employment. I interpreted this statute to require: (1) that any application for change in wages which would ordinarily be the subject of a Form 10 application under the Wage Stabilization Act must be the subject of an application under Section 5 of the War Labor Disputes Act; (2) that any change in contractual relationship with the employees of a seized plant must be the subject of an application under Section 5 of the Act, even though such change did not involve a wage adjustment within the meaning of the Wage Stabilization Act; and (3) that no application might be filed by the management of a company, the plant of which was in possession of a government agency, without the formal authorization of the government agency. The War Labor Board concurred in this interpretation. It was obvious that in as large and varied a group of establishments as the 104 uptown machine shops, occasion frequently arose for the filing of Form 10 application for wage adjustments. In respect of contractual relations, a large number of the 60 unions represented held contracts with one or more of the seized shops; it was obvious that these contracts would be expiring from time to time, and that it would be desired by both the union and the company involved that such contracts be extended, or new contracts negotiated.

When the seizures were originally effected under the Executive Orders, it was not contemplated that Navy Department possession would continue for an extended period. Therefore, it was assumed that it would be unnecessary to effectuate machinery for the making of changes in wages or other terms and conditions of employment which were frozen by the Act. In the event that any company desired to effect such change, it was expected that application therefor would be first submitted to me, in accordance with the provisions of Section 2 of the uniform management agreement providing that all questions of labor relations and policy shall be subject to my direction, and should be reported to me forthwith.

When it became evident that Navy Department possession would continue for an extended period, experience soon indicated that managements could not be relied upon to understand or to carry out their obligations under the War Labor Disputes Act, with which managements were naturally unfamiliar, unless formal machinery was established by me for that purpose. In a number of instances, for example, companies had filed Form 10 applications with the War Labor Board without notice to me, or had entered into negotiations for, or contracts with, some of the many unions represented in the plants. It had been assumed that the War Labor Board, whose action had led to the seizure and whose official duty it was to administer the War Labor Disputes Act, would have called my attention to the filing of unauthorized Form 10 applications involving any of the seized shops. Upon learning that the War Labor Board had overlooked this situation, I directed the attention of the Board to the matter by a letter dated 16 January 1945, in which it was pointed out that no general authorization of mine had been given for the processing of Form 10 applications, and that certain dispute cases were pending of which my office had not been notified. The Regional Board thereupon made arrangements to withhold action upon any application until authorization from me had been obtained.

A further complication developed when the National Disputes Director chose to interpret the provisions of Section 5 of the War Labor Disputes Act as conferring jurisdiction thereunder solely upon the National Board and not upon the Tenth Regional Board. Representations were promptly made by Mr. MacLean to Mr. Berliner, National Disputes Director, as to the adverse effect which would result to Navy operation of the seized plants if every Form 10 application were to be subjected to the delay of being forwarded to Washington and approved by the National Board. At the instance of Mr. Berliner, the National Board, therefore, on 14 March 1944, adopted a resolution delegating its authority under the Act to the Tenth Regional Board, subject to reservation of the usual right to review by the National Board in the event of appeal or on its own motion. This delegation of authority extended only to routine applications for wage or contract adjustment, and not to requests for punitive action against Lodge 68.

Following delegation of authority to the Regional Board, an informal understanding was then worked out: (1) that all applications for which final approval had been given by the Board prior to 1 January 1945 would be regarded as having been filed with my authorization, (2) that applications on which final approval had not yet been given would not be processed to completion until formal letter of authorization had been received from my office, and (3) that no new applications would be considered unless filed with my authorization.

Because of the importance of the policy questions involved, great care was taken in formulating an order to the managements of the seized shops for the purpose of regulating procedures with respect to negotiations for labor union contracts, the method of approval of such contracts, and the procedure for authorization of applications to the Regional War Labor Board. Preliminary draft of the order was furnished to Mr. Berliner, National Disputes Director, for his approval and a number of suggestions made by him were incorporated therein. As promulgated to the managements on 24 March 1945, Order No. 11 set forth the following requirements: (1) Written permission must be obtained from me prior to entering into negotiations with any labor union; (2) Written consent must be obtained from me prior to the execution of any labor union contract by management, or prior to the agreement by management, either oral or written, to any change in the terms or conditions of employment in the seized plants; (3) Upon receiving written consent from me for a proposed change in terms and conditions of employment, application for approval thereof must be filed with the Regional War Labor Board. The order made it clear that grievance procedure was not affected thereby but would continue to be governed by Order No. 8. With the concurrence of Mr. Berliner, the order likewise granted an exemption from its requirements for the filing of applications to the Treasury Department for change in salaries under Treasury jurisdiction and for the putting into effect of changes in wages which ordinarily would be permissible under the general orders or regulations of the War Labor Board without formal request for approval thereof.

The requirement that written consent be obtained prior to the institution of labor negotiations was dictated by the fear that managements might otherwise negotiate for contract provisions of such a nature that I would not consent to their inclusion in a contract. It was felt essential for the maintenance of harmonious relationships with the unions concerned that union business agents be fully aware at the outset of negotiations that they were dealing with companies in possession of the Navy Department and that any contract finally negotiated would require my consent. When managements requested permission to enter into negotiations, they were furnished with a letter of consent which set out the conditions under which any contract might be executed. A copy of this letter was required to be furnished to the union business agents at the outset of negotiations. It may be stated that I proposed to grant consent to enter into negotiations in all cases where a request was made therefor, since to refuse such consent would be in violation of the requirements of the National Labor Relations Act. It is probable that such consent would have been required by the Act even in the case of a request by Lodge 68. As has been previously noted, the request of Lodge 68 of 28 February 1945 was not denied, but officials of the lodge were referred to my office. If the business agents of the lodge had chosen to take the matter up with me, it was intended that all requirements of law be fully complied with.

Under the provisions of Order No. 11, when managements had completed negotiations with labor unions and had presented a tentative contract to me for my approval, I generally authorized the company concerned to execute such contract on its own behalf only subject, however, to the following conditions: (1) that such contract should be presented to and approved by the Regional War Labor Board; and (2) that during the period of Navy Department possession, any provisions in direct conflict with the policy of my office should be deemed suspended. The provisions which were most frequently the subject of such reservation included: (1) provisions authorizing the submission of grievances to third party arbitration. I deemed it inconsistent with the sovereignty of the United States that grievances arising in the seized plants should be submitted to outside arbitration and provided for the substitution of my decision at the final level, in lieu of arbitration; (2) provisions prohibiting the employment of part-time workers. Such provisions were in conflict both with Order No. 7 of mine, previously referred to, and with the policies which should govern in times of labor shortage; (3) provisions prohibiting work in excess of 48 hours per week. In view of the position taken with reference to overtime work by members of Lodge 68, I deemed it necessary and advisable to reserve the right to specify the hours to be worked by members of all other labor unions. In no case did any company management or labor union voice objection to the acceptance of the conditions imposed by me.

I formally advised the War Labor Board that Navy Department authorization for the filing of a Form 10 application for wage increase did not constitute an approval of the increase on the merits, either from the standpoint of wage stabilization or from the standpoint of management policy. The responsibility for wage stabilization was vested by law in the War Labor Board, and I did not deem it necessary to constitute my office as an additional Board of Review. With regard to management policy, the uniform management agreements imposed responsibility for profitable operations upon the civilian managements, and I did not desire to interpose any unnecessary interference therewith. It was not felt, however, that I could close my eyes to the contents of applications for increased wages in all respects. Prior to granting authorization for any application, the facts therein stated were reviewed by the staff. In those few instances where positive misstatements of fact were found, the applications were returned to management with instructions to correct the same. In other instances, where applications asked rates which appeared to be entirely excessive, it was considered proper to discuss the matter with the civilian managements to determine whether they would voluntarily revise such applications. If management insisted upon filing the application, such filing was authorized, but by letter of transmittal from me the attention of the War Labor Board was called to any fact which appeared properly subject to comment, such as the fact that the proposed rates were substantially in excess of those paid by comparable companies in Navy Department possession.

A complicating factor with respect to negotiations for and the execution of labor union contracts was the existence in San Francisco of a large number of "employers associations" of which some or all of the seized shops were members, such employers associations being the customary bargaining agents for the employers in the case of labor contracts. I deemed it advisable to interpose a minimum of interference with the normal relationship between the seized shops and the employers associations of which they were members. However, it was equally inadvisable to permit any relaxation of the provisions of Order No. 11 merely because of the membership of the seized shops in such associations. Therefore, I generally permitted the various employers associations to continue to represent the seized shops in negotiations, upon the definite understanding that my consent was equally necessary in such cases, either for the initiation of negotiations or for the final execution of any contract which might be binding upon one or more of the seized shops. All employers associations with which my office had contact were most cooperative and voiced no objection to the supervision which I necessarily imposed over certain of their members.

On 10 April 1945, I was relieved as Officer-in-Charge by Commander H. K. Clark, USNR.

At the end of World War II, the 104 uptown machine shops were turned back to their owners, and shortly after that the American Federation of Labor moved in, took possession of Lodge 68, and dispensed with Messrs. Hook and Dillon.

For the successful operation of the 104 uptown machine shops in San Francisco, great credit is due to my counsel, the indefatigable Charles C. McLean, Jr., to the especially competent management of Commander H. K. Clark, USNR, and to Captain Harvey S. Haislip, USN, Assistant Chief of Staff of the 12th Naval District, who kept my relations smooth with the District Organization.

Seizure of the Plant of the Lord Manufacturing Company at Erie, Pennsylvania

Pursuant to Executive Order No. 9493, I seized the plant of the Lord Manufacturing Company at Erie, Pennsylvania, on 25 October 1944 and immediately set up an organization or staff to operate this plant consisting of myself, a resident officer-in-charge, a counsel, a comptroller, a disbursing officer, and a cost inspection officer.

Effective 1 November 1944, I entered into an agreement with the General Electric Company by which that company furnished the personnel to function as my managing agent. Specifically this function was designed to cover all phases of management except (a) setting of sales prices, and (b) public and labor relations.

Mr. R. C. Reed of the Erie works of the General Electric Company became the manager, and Mr. D. A. Smith, the Chief Accountant. Mr. Herman Emmett, the manager of the Erie Works, was ceaseless in his attendance at the plant and at plant conferences and in his interest in the entire operation. This management was procured as a result of my personal intervention with Mr. Charles E. Wilson, President of the General Electric Company, who could never say no to an appeal from a government official for help in time of war.

The seizure of the plant and facilities was made without any interruption of any kind in the normal production, sales, purchasing and other activities. There was no stoppage of work due to a labor dispute, and at the time of turnback there were no unsettled grievances. The plant was operated with Federal money throughout the occupation.

The Lord Manufacturing Company was engaged in the production of shock mountings for the war effort. These mountings—a combination of metal and rubber, properly bonded—reduced to a satisfactory degree the amount of vibration transmitted for instance from an airplane engine to a plane, or from the plane to the instruments on the switchboard. This principle has many other applications. It is interesting to note some of the unique problems that wartime accelerated development placed upon this type of industry.

There was a shortage of crude rubber.

Tremendously increased torque loads were imposed upon aircraft mountings because each conventional radial aircraft engine was "souped up" to deliver far beyond its original rated horsepower.

High-altitude flying was attended by extremely low temperatures and an increased amount of ultra-violet light and nothing but crude rubber would stand up under these severe conditions.

Shock devices were manufactured by other industrial concerns as well, and the Army and Navy felt that the prices charged to the Army and Navy by the Lord Manufacturing

Company in comparison were such as to warrant action by the Armed Services.

Accordingly, on 2 October 1944 the Army and Navy Joint Repricing Board served Repricing Order No. 1 on the Lord Manufacturing Company. This order required that on and after the above date, sales, contracts, and undelivered balances of previously executed sales contracts were to be made at prices fixed in the repricing order. In general this order instituted a new price schedule for only a portion of the items manufactured by this company since it referred only to items produced by the Lord Manufacturing Company under contract with the United States Government or a prime contractor of the government.

During the period from the date of the repricing order, which fixed fair and reasonable prices, to the time of occupancy, the Lord Manufacturing Company was alleged to have followed a policy of requiring customers to make payment in advance or on delivery of shipments, and to have listed on each invoice the selling prices before revision as well as prices required by the repricing order. The company is also alleged to have typed the following notation on each invoice: "Shipments . . . during the period from October 2, 1944 through March 31, 1945, are subject to Army-Navy Joint Repricing Order No. 1. This Order fixes the prices of these articles (to which we do not agree). The action of the Government with respect to repricing compels us to require cash prior to shipment; kindly arrange to send us your check to cover your open balances."

The Army and Navy consequently felt that the Lord Manufacturing Company had failed to comply with the terms and conditions of Army and Navy Joint Repricing Order No. 1, fixing fair and reasonable prices for essential war materials. This was the first and to the best of my knowledge the only instance where a plant was seized during World War II for the reasons stated.

The major part of the vibration or shock mountings went to the Air Force—about 75%—but the Navy got stuck with the job. The Army merely washed their hands and said they had no one to assign to this project.

It may be interesting to some to quote the law behind this repricing and seizure. The Revenue Act of 1943, Section 801, paragraph (d), provides:

"Whenever any person willfully refuses or willfully fails to furnish any such articles or services at the price fixed by an order of the Secretary in accordance with this section, the President shall have power to take immediate possession of the plant or plants of such person and to operate them in accordance with Section 9 of the Selective Training and Service Act of 1940, as amended."

The Selective Service Act states:

". . . the President through the head of the War or Navy Department of the Government, in addition to the present authorized methods of purchase or procurement, is hereby authorized to take immediate possession of any such plant or plants, and through the appropriate branch, bureau, of the department of the Army or Navy to manufacture therein such product or material as may be required, and any individual, firm, company, association, or corporation, or organized manufacturing industry, or the responsible head or heads thereof failing to comply with the provisions of this section shall be deemed guilty of a felony, and upon conviction shall be punished by imprisonment for not more than three years and a fine not exceeding $50,000.

"The compensation to be paid to any individual, firm, company, association, corporation, or organized manufacturing industry for its product or material, or as rental for use of any manufacturing plant while used by the United States, shall be fair and just. . . ."

The company not only did not have adequate cost accounting returns, but cost records were not available. Books of account were some eight months behind at the time of seizure. One of our first tasks was to bring the books of account and records up to 2 October 1944. I questioned the data on which the Repricing Board had based its new prices and I was frankly faced with the proposition of making these new prices produce the necessary profits to justify the Government's action in assigning new prices. I had had nothing to do with this controversy. I had had nothing to do with the Government's price fixing, and as usual I was thrust into the breach to pull the Government out of a hole.

As in so many other seizures, the motives of those who precipitated the seizure and my motives in trying to protect the United States Government from loss in Court Action were entirely different. My job, although never even hinted to me in any way by those groups who precipitated seizure, was not only to continue or even accelerate production, but also to do all those things which would fortify the United States against suit for damages. This was an idea of my own. In all my 110 seizures under Executive Orders, no one ever suggested this attitude to me.

Even the most superficial acquaintance with the Lord Manufacturing Company suggested that the introduction of some of the tried and proven methods of industrial management so common, so well known in Detroit, Pittsburgh, and Schenectady, could easily justify, from the profit standpoint, the arbitrary price list determined by the Army and Navy Repricing Board, which in my opinion had no more justification in fact than the Lord prices themselves.

In order to insure profits at reduced prices, a survey was made of existing conditions which resulted in the following decisions:

(a) A delegation of responsibility; there was too much congestion at the top.

(b) Installation of a standard cost system instead of a job order cost system. A job order cost system is appropriate where large sums are spent on a single job such as building a ship, or power plant machinery, but inappropriate when production consists of a large volume of relatively small priced articles—be they shock mountings or safety razor blades. Current cost information was not available to either management or the production department.

(c) Provision for closer coordination between the production departments' requirements of material and the purchasing of such material.

(d) Replacement of production schedules which were more historical than useful by schedules which were up to the minute and with which compliance was expected.

(e) Establishment of a budget to control costs and create an incentive to meet predetermined standards.

(f) Promulgation of rules and regulations covering the conditions under which overtime might be authorized for the guidance of supervisors and department heads and specific instructions to the payroll department concerning the proper method of calculating overtime allowances, permissible under the law.

(g) Installation of regular meetings of the department heads in order that current problems of plant management could be considered and policies could be formulated.

(h) Reexamination of the company's bonus plan on the basis of ability to pay.

(i) Promulgation of plant instructions on general subjects, as well as detailed operating instructions.

(j) Establishment of an active safety program.

(k) Classification of the procedures to be followed so that union representatives and foremen could facilitate the handling of grievances.

(l) Study and review of rate classification of jobs in order to produce compliance with accepted standards.

(m) Development of the inspection systems to improve methods of reporting defective workmanship and material in order to bring about an improvement and correction, not only on the part of suppliers of material, but also on the part of shop employees through a spoilage and waste campaign.

(n) Improvement in handling material to save time and space.

(o) Promulgation of written instructions regarding the laws of the State of Pennsylvania in regard to the employment of females.

(p) Strengthening of the authority of the personnel department.

(q) Emphasis on the urgent necessity for expediting.

(r) Reduction where possible of the infinite number of designs in order to lessen a difficult manufacturing problem.

(s) Setting up of a Small Order Section to handle special items, thus eliminating interruption in standard production.

From an accounting point of view, the operation at Lord's is a beautiful example of two principles: (1) accounting is the most powerful tool in the hands of the manager in controlling costs and labor; (2) an accounting system must take consideration of the item price and volume of item production. In other words, accounting must be adapted to the prevailing product.

Now "shop cost" depends on: (a) expense for direct labor, (b) expense for direct material, (c) manufacturing overhead, (d) manufacturing losses. If we add two more "costs" we will have the definition of "manufacturing cost." These two items are: (e) the expenses of the engineering department, and (f) the expenses of shipping, packing, and boxing. But before we price the article to be sold we must add (g) the cost of general and administrative expense, and (h) allowance for profit.

For the class of production engaged in by the Lord Manufacturing Company it was the judgment of the General Electric Company that "cost control through the use of cost standards" could be "accomplished through analysis of variance" from reasonably established standards. To quote from Mr. D. A. Smith, our General Electric Company loaned expert on accounting, "cost control through use of cost standards is accomplished through analysis of variance. [That is, deviation from set standards.] This might be termed as accounting for exceptions. In my opinion it is much more desirable to devote time to cost analysis work of the exceptions rather than to spend excessive time on accumulating costs which a job order cost system demands. Through separate analysis in accounting for these exceptions, current control may be exercised, and in my opinion, this is vastly more important than a mere writing of historical costs;

also, [I] believe that historical costs are not overly useful in price determinations, since products are priced and sold by the time costs are generally known. If prices are based on cost standards, and effort is directed to meeting these standards, profits of course should be realized."

The services of the American Appraisal Company were secured to appraise the value of the land, buildings, machinery and equipment, dies, jigs, fixtures, molds and mold caps, raw materials, work in process, component parts, finished merchandise, supplies and spare parts.

The services of Price, Waterhouse and Company were secured to review and report on the financial statements and records of the Lord Manufacturing Company.

The services of Mr. John Berry, a Certified Public Accountant of Buffalo, New York, were secured to: (a) price at cost or market, whichever was lower, physical inventories of the Lord Manufacturing Company; (b) prepare the journal entries necessary to record all such physical inventories of the company and make any and all postings, entries and adjustments of records and accounts necessary to accurately reflect said inventories on the books and records of the company as of 30 September 1944; (c) supervise and furnish assistance, in connection with the posting of the general and factory ledgers and any subsidiary records for specified periods and compare and adjust the inventory accounts shown by the factory ledger as of 24 October 1944, with amounts as determined from the perpetual inventory records.

So much work was done in the complete reorganization of the Lord Manufacturing Company, and the records are so voluminous that it is impossible to convey more than a general idea of the high points. The records themselves are an education in industrial management.

The greatest achievement was the installation of the cost-variance system which was accepted by the organization as a tremendous improvement over the job costing method. It provided that estimates could be easily prepared for price determination. It furnished currently cost control information as follows:

(1) Direct labor efficiency was reported weekly.

(2) Indirect labor efficiency was reported weekly.

(3) Manufacturing losses were summarized and reported monthly.

(4) Report of variance for purchased material was issued monthly.

(5) Operating information was accumulated which would permit installation of a flexible budget for overhead control.

(6) Operating results became known monthly.

By 1 April 1945 it appeared that savings in the cost of operation had greatly exceeded a million and a quarter dollars and that there would be no difficulty in maintaining profits at a satisfactory level. On this date a new agreement was entered into with the General Electric Company whereby the Officer-in-Charge could requisition assistance as required from that company on a basis of part-time employment. A new organization had been put into operation and the necessity of further help from the General Electric experts no longer existed. The plant had been given a system that money couldn't buy.

The Lord Manufacturing Company eventually sued the Government, and the case was settled out of court. At any rate, the Government approached the settlement with a large sum in the U.S. Treasury, representing profits.

The Navy Department wanted me to set up a patent system with branches in some of the principal cities of the United States. Eventually this activity developed into the Office of Research and Inventions and finally the Office of Naval Research. Accordingly, on 10 April 1945 I was relieved by Commander Harry K. Clark, USNR, in private life a manufacturer of wide experience.

Captain Harvey Collison, who was the resident Officer-in-Charge during the entire operation, worked ceaselessly and without stint on all reorganization problems. Mr. Frank W. Marshall, special cost accountant in the Bureau of Supplies and Accounts, also contributed largely to the success of the operations. Mr. Marshall had been with me in practically all my seizures, had worked ceaselessly for me from August 1941 to April 1945, and I remember gratefully his unlimited contributions to our success in this original and unusual line of work.

6

Office of Naval Research

I HAD long been interested in the unsatisfactory patent situation in the Navy and took every opportunity to interest Mr. Forrestal in its improvement. The Navy had a Patent Section which was located in the office of the Judge Advocate General of the Navy. However, it did not have any field representatives in the Bureaus, Laboratories, or Navy Yards. And, above all, it was dead from the neck up.

While I was Director of the Naval Research Laboratory, as previously indicated, I obtained authority from the office of the Judge Advocate General to establish my own patent office at the Laboratory and, for the head of this office, I selected a young Reserve officer, Lieutenant Commander Newell A. Atwood, who was both a radio engineer and a patent lawyer. There had been a great deal of criticism in Congress following World War I about the way the Navy had handled, or rather had not handled, its radio patent situation. It was alleged that many radio inventions financed by public funds had slipped through the fingers of the Navy Department and into commercial hands. I realized that I must protect the Government's interests in this respect, especially in radar inventions.

A practice was instituted at the Laboratory, under my guidance, requiring a running inspection of the scientists' notebooks by my patent lawyer to see that inventions were properly described by the scientist concerned so that the record would stand up in court in infringement cases. I also required the photostating of all notebooks once a week and the deposit of the original notebooks in the fireproof vault each night and during holidays and weekends. The scientists didn't like this procedure a bit, but they were quieted down by Dr. A. Hoyt Taylor who so often was used to pour oil on the troubled waters. Dr. Taylor had a very broad point of view of Naval administration due to his long Government service as a civilian scientist and Naval Reserve officer and to much service aboard ship in the Fleet.

On one occasion I accompanied Forrestal at hearings before a Senate Committee on patents. Forrestal and I were defenders of the patent law as then constituted and practiced in the Naval Establishment. When I became Director of the Naval Research Laboratory, I changed the official attitude toward patents. I made it a duty to apply for patents and removed the stigma which had formerly been thrown around the persons of those who applied for patents. The official attitude had been that scientists who applied for patents were "commercially minded" and not truly interested in Government objectives.

The patents were prosecuted at Government expense, the inventor retained his commercial rights, and the Government owned the patent otherwise. The retention of commercial rights was not as much of a concession as it might seem because important inventions were kept secret for years, as the Government is privileged to do by law. The commercial rights arrangement was an incentive, however, and did serve as a reward for diligence and creative thinking. Later the Government's attitude toward commercial rights in the case of inventions made on Government time became the subject of an investigation by the Department of Justice, a matter which I shall discuss later.

On 2 October 1943 Mr. Forrestal, then Under Secretary, requested Mr. R. J. Dearborn, President of the Texaco Development Corporation and Chairman of the Committee on Patents of the National Association of Manufacturers, to survey the policies and procedures of the Navy Department with respect to patents and inventions. A happier choice could not have been made. Mr. Dearborn made the survey and submitted a report on 10 March 1944, together with the plan which became the basis for the reorganization of the Navy Department activities concerned with research, inventions and related matters. This report was accompanied by a draft under which an Office of Patents and Research would be established. Actually Mr. Dearborn submitted three plans, but it was desired to proceed cautiously in order not to stir up too much opposition.

Office of Patents and Inventions. Many of the proposals set forth in the third plan of the Dearborn report were embodied in a directive issued by the Secretary of the Navy on 19 October 1944, under which there was established an Office of Patents and Inventions. This new office was charged only with certain functions concerning inventions and patents, and it was given no authority over the research activities of the Navy Department. I became the head of this new office upon the recommendation of Mr. Dearborn and was, at the same time, Chairman of the Patent Royalty Revision Board.

While a small step and only a beginning, it was a fundamental step in that it took all responsibilities for patents from the office of the Judge Advocate General and placed them in a patent office under aggressive leadership.

Patents. I immediately went to work on a "grass roots" patent system, something the Navy never had had. I worked especially on the Navy Yards because the Navy buys the latest equipment and sooner or later it ends up in a Navy Yard for repairs or alterations. Being the latest equipment, it is almost sure to be susceptible to improvement. After a first-class artisan or instrument maker has overhauled a few hundred or more, he has definite ideas about how the equipment can be improved. If the Government doesn't patent the idea, the manufacturer will, naturally.

After the new arrangements were explained, the repairmen couldn't believe that when they had a new idea—an invention—all they had to do was to call up the nearest branch patent office and a patent lawyer would come right down to look into it. No drafting or writing of a patent application—the branch office handled all of that.

Arrangements were also made to service inventions from the Fleet. Branch offices were established in all Bureaus.

As a result of this drive on patents, the Navy was soon filing as many patent applications as several large corporations combined. During slightly more than a twelvemonth, 2,516 disclosures were received and 2,343 patent applications filed. In possible extenuation, it must be said, however, that commercial concerns were undoubtedly delayed in their patent work by the fact that a large number of their patent lawyers were attached to the armed services.

After the war it was noticeable with what success the Naval Research Laboratory had defended the Navy's claim to radar patents[1]—quite a different situation from that which the Navy faced in radio following World War I.

Commander Ralph Chappell, USNR, toured the country with me organizing branch patent offices. His assistance was invaluable in this pioneer work. He headed the patent section in the Office of Patents and Inventions and later in the Office of Research and Inventions until relieved by Captain John H. Austin, USNR.

Patent Royalty Revision. When royalty for use of patents was reported as an item of cost on Navy procurement contracts, an investigation was made to determine whether, under wartime conditions of procurement, the aggregate royalties to be paid by the Navy were reasonable. If such investigation revealed that royalties were excessive, steps were taken to reduce them, either by negotiation and agreement with the recipient of the royalties or by an order issued by the Secretary of the Navy. If agreement could not be reached, an order was issued to the party or parties paying the royalties not to pay more than a specified percentage royalty. Usually a maximum annual sum for the royalties was fixed. Any excess of royalties required by agreement to

[1] See Chapter IV: Naval Research Laboratory, p. 180.

be paid above the percentage or annual fixed maximum was recapturable by the Government.

The huge procurement incident to war caused royalties to assume astronomical proportions. The purpose of the law was to allow those to whom royalties were paid to receive reasonable amounts but not to allow them to receive excessive profits due exclusively to the exigency of war. The law was exceedingly well written and was easy to administer. No cases handled by the Royalty Revision Board were ever appealed in court, but at the instance of the Department of Justice the Board assumed jurisdiction in several cases originating in boards of other armed services and succeeded in keeping them out of court by reasonable and informed negotiation.

Including cases involving joint Army and Navy action, it was estimated that over 21 million dollars was saved the Government with an expected saving of over 13 million dollars more. Much of the success of this Royalty Revision Board was due to the energy and skill of Captain Robert Whiting, USNR.

Justice Investigates Government Patents. All Executive Departments were investigated during this period by the Justice Department to ascertain patent practices and to formulate a recommendation for a uniform practice. For instance, the Naval Research Laboratory, as well as the Navy in general, allowed Government employees who made inventions on Government time to retain their commercial rights. The Bureau of Standards, under the same circumstances, did not allow its employees to retain commercial rights. What was the result? The U.S. Patent Office called my attention, unofficially of course, to the fact that, although the Research Laboratory was a much smaller institution, it submitted far more patent applications than did the Bureau of Standards.

I argued that hope of commercial rights acted as an incentive. My opponents in Congress argued that, if a money incentive were necessary, it indicated that *all* the Government scientists should have their pay raised. It seems to me that that is pure Socialism. Anyhow, the object of a research laboratory is to make inventions and discoveries, period. Policy makers should not interfere with the *management* of Government activities.

After long delay, the report of Justice and the recommendation to deprive Government workers of commercial rights was published and ordered. I feel that the decision is wholly wrong and that it is characteristic of a certain section of society which has had absolutely no experience in getting work out of people.[2]

Office of Research and Inventions. On 19 May 1945 the Secretary of the Navy issued a directive establishing, in his office, the Office of Research and Inventions. This directive, following the recommendation of the Dearborn report, provided for the appointment of a Chief of Research and Inventions to head the new office. I was appointed to that new office on the same date as the directive was issued.

Not until the Secretary issued the directive had any attempt been made to coordinate and centralize the research activities of the Navy, except for my abortive attempt before Pearl Harbor, as previously noted.

I was given broad powers with respect to coordination and conduct of Naval research activities and administration of activities relating to patents and inventions. The directive also provided for the incorporation, under the Chief of Research and Inventions, of the Office of Research and Development (from the Office of the Secretary), the Naval Research Laboratory (from the Bureau of Ships), and the Division of Special Devices (from the Bureau of Aeronautics). Another directive, dated 4 February 1946, set forth the procedure to be followed by the Bureaus so that the negotiation and execution of Naval research contracts

[2] Government employees were deprived of all patent rights by Executive Order 10096, dated 23 January 1950. Some patent lawyers doubt the legality of this Executive Order. See Journal of the Patent Office Society, Vol. 35, No. 6, pp. 426-434.

would be coordinated by the Office of Research and Inventions. On 29 September 1945 the President issued an Executive Order, No. 9653, under authority of the First War Powers Act, in order to eliminate any possible doubts concerning the authority of the Secretary of the Navy to make effective the organization changes referred to above.

On 1 August 1946 Public Law 588 of the 79th Congress created the Office of Naval Research in the Office of the Secretary of the Navy, with Bureau rank, and I became its first Chief, by and with the advice and consent of the Senate. Thus the Office of Research and Inventions, under a new name, acquired statutory authority and its own appropriation and my dreams for Naval research were realized. In Washington, one cannot accomplish anything without a statute in front of him and an appropriation behind him.

There were some interesting highlights in connection with the passage of this legislation. First, hearings were held on a bill introduced in the House of Representatives to establish an Office of Naval Research in the committee room of the House Naval Affairs Committee, the Honorable Carl Vinson, Chairman. The bill as introduced was most comprehensive and would have centralized the control of all Naval research in the Office of Naval Research. Both Mr. Vinson and Mr. W. Sterling Cole, the minority leader, had very decided views on the subject. Mr. Vinson excoriated the Navy Department for its continued neglect in not setting up such a center long before. This was music to my ears but, as I was the senior representative of the Navy Department during all of this hearing, the way Mr. Vinson ripped the Navy Department up the back for the alleged mishandling, or lack of handling, of research over the years made the attacks feel almost personal. I took it all smilingly, however, because what does a little misplaced contumely amount to when one is getting what he wants? Also, the fact that I, who had introduced so many new ideas into the Navy Department, should be the one who had to bear the brunt of Mr. Vinson's accumulated resentment against the scientific setup in the Navy Department struck me as quite funny. Every time I looked at the Committee members and saw the smiling face of my friend, Representative Jack Z. Anderson of California, I realized that those who knew also recognized the humor of the situation.

The Senate hearings were quite different. I was not the Senior member from the Navy Department. I was flanked by Mr. W. John Kenney, Under Secretary of the Navy, and representatives from Naval Operations and the Bureaus. The House bill was emasculated to a certain extent. Coordination of Bureau research was substituted for control, but so many good features of the original bill, such as locating the Office of Naval Research under the civilian Secretary and not under the Office of Naval Operations, were retained that it would have been folly to have interfered. I recognized that the Navy Department had been pushed as far in this matter as it could be pushed in my time.

The foundations of the Office of Naval Research were laid during the existence of the Office of Research and Inventions, 19 May 1945 to 29 September 1945, and an examination of this period is necessary to understand the origin of the policies and practices of the Office of Naval Research.

Founders. I have already mentioned the name of Mr. Dearborn as the initiator of the plans which resulted in a chain of events ending in the statutory establishment of a research center in the Office of the Secretary of the Navy. Others without whom this achievement could not have been accomplished were Rear Admiral Lewis L. Strauss, USNR, and the Honorable H. Struve Hensel, Assistant Secretary of the Navy under Mr. Forrestal. Great credit is also due to Rear Admiral Luis deFlorez, whose Division of Special Devices of the Bureau of Aeronautics became a part of the new Office of Research and Inventions. He contributed many of the key individuals of Special Devices to the main office of the new organization, and he obtained from the Bureau of Aeronautics a transfer of funds from that Bureau to the new organization amounting to $24,650,000.

Organization. At the very beginning I had some fundamental decisions to make regarding personnel and contracts. For years naval officers and scientists had worked together harmoniously and efficiently at the Naval Research Laboratory, and I decided to use the same system in my new office. I selected all my division heads from among the scientists. I assigned Naval officers to all administrative positions. I procured the services of Dr. Alan T. Waterman of Yale University as the chief of all our scientists and he did an excellent job, not only in science but in the necessary coordination work between science and administration.

The work on contracts for research was original and pioneering. In usual Government contracts one can see what is bought—it is tangible, the quality and quantity have to be as described in the specifications and, if machinery, it works or it doesn't. Before the contract is sent to the Comptroller General of the United States to validate the payment, it must bear a certificate that value has been received.

Now we must let contracts for the intangible—research. Obviously there could not be any specifications. Results or end products would be opinions, discoveries and inventions, possibly, or even reports that the problem had reached no conclusions and the work should be discontinued. The necessary type of contract was conceived by the late Captain Robert D. Conrad, USN, Commander James H. Wakelin, USNR, and Commander John T. Burwell, USNR, as a result of their experience with contracts in the Office of the Coordinator of Research and Development before it was absorbed by the Office of Research and Inventions. They were assisted by Commander H. Gordon Dyke, USNR, Commander Eugene Krause, USNR, and Commander Bruce Old, USNR.

Captain Conrad was the administrative head of the Scientific Research Division. A great deal of the original success of the Office of Naval Research was due to his extremely likable disposition coupled with an active and brilliant mind. His death from natural causes at an early age was a great loss to the Navy.

Forty-eight of these contracts were shortly let with universities and non-profit institutions and twenty more with commercial corporations.

It was I who had to take the responsibility for letting these contracts which were not provided for by law. I believed that the emergency justified the action, but I was careful when the opportunity presented itself to tell Congress, the Bureau of the Budget, and the Comptroller just what I was doing so that they would get the news first—a very practical approach to a problem in Washington.

I did have trouble explaining to the Budget and the Appropriations Committee, when I had hearings on my 1946 budget, what research was. I explained in response to questioning that, if you could write specifications for a research problem, there was then no research, it was engineering. In an attempted witticism I told them that if you knew what you were doing it wasn't research.

As a result of all this, the law which established the Office of Naval Research contained a provision authorizing the Secretary of the Navy and, by his direction, the Chief of Naval Research and the Chiefs of the Bureaus to enter into contracts and amendments and modifications of contracts for services and materials necessary for the working and securing of reports, tests, models, apparatus, and for the conduct of research without regard to certain statutory limitations, including section 3709 of the Revised Statutes, which might otherwise be applicable. This incident further confirmed my experience that, if you know what you want and can explain it properly, you will have no trouble getting it from Congress provided, of course, that your request is reasonable.

The law also provided for an Advisory Committee to the Office of Naval Research, and we were fortunate in being able to obtain the services of Dr. Karl T. Compton as the first Chairman.

Branch Offices. The Special Devices Division of the Office of Naval Research long had had branch offices in Boston, New York, Chicago, San Francisco, and Los Angeles. They were engaged primarily in synthetic training methods and also acted as a reservoir for spare parts for such training devices. When the Special Devices Division joined the Office of Naval Research, I merged the branch patent offices with the Special Devices branch offices to form the new branch offices of the Office of Naval Research.

Other duties of the branch offices were to keep in touch with research contracts in their respective areas as well as progress in research. A branch office was established in London for the purpose of becoming familiar with the general course of research in England and on the continent. It did not handle secret matters; it was not a cloak and dagger club.

Staff Organization. I was assisted by an administrative branch under Captain Martin J. Lawrence, USN, which planned, coordinated, and controlled general administration and management services; a material branch under Commander Mason which drew up all research contracts (the patent clauses of procurement and research and development contracts were skillfully handled by Commander Ernest F. Vanderstucken, USNR); a budget branch under Lieutenant Commander C. H. Helsper, USNR, which was responsible for proper and legal expenditures of research funds, for avoidance of any deficits, for compliance with statutory limitations and restrictions on the use of research funds and material, and for the preparation of the budget estimates; a field service section under Lieutenant Commander Paul Busse, USNR, which supervised administrative and operational procedures of branch offices; the office of the General Patent Counsel for the Navy under Captain John H. Austin, USNR, which was a patents section; an inventions section under Dr. Murray O. Hayes which analyzed inventions submitted by Navy personnel and others; a publications branch under Mr. Earl Nash which furnished information regarding the Navy's research programs and activities; a public works facilities section which was responsible for matters in the Office of Research and Inventions pertaining to real estate, leases, and construction projects; a scientific personnel branch under

Mr. A. H. Hausrath which was concerned with activities pertaining to attracting, developing, and retaining highly creative and productive scientists and technicians in the Naval Research Laboratory's establishments; personnel administration (military and civilian) under Commander G. L. Calloway, USNR; and a planning division under Captain Conrad which was responsible for planning, recommending the placement of, and directing contracts for fundamental research in civilian and Government establishments and for placing and coordinating research within the Naval Establishment.

Captain Martin J. Lawrence, formerly shop superintendent and later Executive Officer of the Naval Research Laboratory, was the Executive Officer of the Office of Naval Research and, specifically, in charge of all contracts. With his Vermont appreciation of the fact that money is the leakiest substance in the world, he performed his contractual duties as "Johnny at the rat hole," with distinction and to my entire satisfaction.

I never saw an Office get formed and underway so quickly. Captain Lawrence reported for duty on the same day he was detached from the Naval Research Laboratory. He was ably assisted in creating an organization by Commander Virgil Wiese, USNR, a dynamic and irrepressible doer. Many of these very capable Reserve officers came from the former Special Devices Division.

The Office of Naval Research was baptized at the National Meeting of the American Association for the Advancement of Science at St. Louis in the summer of 1946. I read a paper before the Physics Section. For the first time scientists realized that, by statute, the Navy Department was in a position to recognize their profession.

One of the greatest achievements of the Office of Naval Research was the establishment of the Scientific Personnel Division under Mr. Hausrath. When demobilization took place, the Office of Naval Research had the only list of young scientists who had been drafted, were now being discharged, and would be available for Government laboratories and industry. By that time the list contained 2,000 names, and professors, directors of laboratories, and industrial organizations constantly availed themselves of this list and often interviewed prospective candidates.

Office of Scientific Research and Development. The Office of Research and Inventions, fortunately, was able to take over and keep alive many problems from the Office of Scientific Research and Development, upon the closure of that office, which otherwise would have expired.

Manhattan Project. The Atomic Energy Act became law after the establishment by law of the Office of Naval Research. There was nothing in the language of the two acts which should have prevented the Office of Naval Research, its scientists and top officials, from being accredited to Manhattan, but there again was the same old iron curtain. After fruitless efforts I finally appealed to our Under Secretary, W. John Kenney, to see what he could do to break the barrier. It is to Mr. Kenney's everlasting credit that he succeeded. My date of retirement from the Navy was so imminent that I considered it good judgment not to avail myself of the privilege of exercising my credentials.

National Inventors' Council. On July 11, 1940, the Secretary of Commerce, then Harry Hopkins, invited Charles F. Kettering to form the National Inventors' Council. It was my good fortune to be the Navy representative on this council from its inception until my retirement. It was a remarkable group of men; each one had an amazing record of accomplishment. The original members were: Charles F. Kettering, Chairman, Lawrence Langner, Secretary, George Baekeland, George W. Codrington, Conway P. Coe, Dr. William D. Coolidge, Watson Davis, Dr. Frederick M. Feiker, Dr. Webster N. Jones, Dr. Fin Sparre, Orville Wright, Dr. George Lewis, and Frederick N. Zeder.

The record of the Council is impressive. Over 210,000 inventions and suggestions were evaluated, and 11,040 cases were forwarded to the armed services as having merit. Over

100 of these inventions saw war service, and tests on hundreds of others were in progress at the cessation of hostilities.[8]

Whether justified or not, the armed services have the reputation for giving civilian inventors the brush-off. Such a body as the National Inventors' Council, by its very existence, becomes a source of encouragement and confidence to inventors. The activities of the Council relieved the armed services of a tremendous amount of work. John C. Green, of the Department of Commerce, acted as Chief Engineer of the staff to the satisfaction of all.

There is one aspect of this contact of the armed services with industry, patents, and technical education of which we should not lose sight. The armed services cannot know too much about these matters, and the association of their representatives with the distinguished members of the Council, I am sure, exercised a most broadening influence.

Joint Office of Research and Development. It had long been apparent to me that the Army, Navy and Air Force couldn't go on indefinitely asking Congress for money for research and development, merger or no merger. There was too much overlapping. General Walter A. Borden, USA, agreed with me, and we started to convince our respective services that something ought to be done. General Borden may be described as having a job in the Army somewhat resembling mine in the Navy. His activity was under the General Staff of the Army but my office was under the Secretary of the Navy by my arrangement. He wanted the Joint Board under the Joint Chiefs of Staff while I wanted it under the civilian Secretaries jointly.

Assistant Secretary of the Navy Hensel interested himself in the matter, with the result that it was agreed that it should be under the civilian Secretaries. Mr. Hensel returned to civilian life and was relieved by Mr. W. John Kenney. Lo and behold, Mr. Kenney discovered soon after he took office as Assistant Secretary that Mr. Forrestal had signed an

[8] *The Magic Curtain*, by Lawrence Langner, E. P. Dutton & Co., New York, 1951, p. 366.

order agreeing to place it under the Joint Chiefs of Staff. Mr. Kenney got busy, persuaded Mr. Forrestal to unsign his agreement, and the Joint Research and Development Board was finally placed under the civilian Secretaries and, after the merger was accomplished, under the Secretary of Defense.

About the time I became Chief of the Office of Naval Research, I was appointed a Vice Admiral as a result of having been recommended for that rank for a third time by the Bureau of Naval Personnel. It took a Denfield, a Nimitz and a Forrestal, in conjunction, to put me over, not to mention such backer-uppers as Strauss, Kenney, and Hensel.

Mr. Forrestal obtained for me a Distinguished Service Medal with the following citation:

"For exceptionally meritorious service to the Government of the United States in duties of great responsibility as Chief of the Naval Research Laboratory, Special Assistant to the Secretary of the Navy, and as Chief of the Office of Research and Inventions, throughout World War II. Responsible in large degree for the development and installation on United States Naval vessels of high-pressure, high-temperature propulsion machinery, Rear Admiral Bowen also exerted strong influence in the development and subsequent adoption by the Navy of radar equipment and other new weapons of radical design, thereby increasing the operational efficiency of the Fleet. Diplomatic and resourceful in his relations with both management and industry, he rendered invaluable assistance to the Secretary of the Navy during critical production emergencies in taking over the operation of plants manufacturing Naval munitions and, by his quick decisions and firm administration of such plants, materially furthered the production of war material for the Navy. Undertaking the organization and establishment of the Office of Research and Inventions during a period of unprecedented difficulty and stress, he worked tirelessly in building a sound foundation for continued technical progress in a field of vital importance to the Navy Department. A leader of rare insight and judgment, Rear Admiral Bowen achieved notable results in these several and varied activities which contributed directly to the successful prosecution of the war, and his vision, initiative and constructive efforts throughout reflect the highest credit upon himself and upon the United States Naval Service."

It was a long hard battle getting vindicated, but it was well worth it.

Not long after this I was detached as Chief of the Office of Naval Research and, on 1 June 1947, retired from the Navy, having reached the statutory retirement age and after having served in the Navy for 46 years lacking 7 days.

APPENDICES

A. Fig. 1. Variation of Efficiency with Temperature and Pressure.

B. Fig. 2. Variation of Theoretical Steam Rate with Temperature and Pressure.

C. Fig. 3. Comparative Economy of Destroyers.

D. Fig. 4. Comparative Economy of Cruisers.

E. Battleship Advisory Board.

F. Engineering Data.

G. Achievements of the Bureau of Engineering.

H. Letter from Charles Edison, Secretary of the Navy, to the Honorable James G. Scrugham, under date 13 February 1940.

I. Navy Department press release, 13 June 1943.

J. Difficulties in Introducing Electric Drive in the Navy.

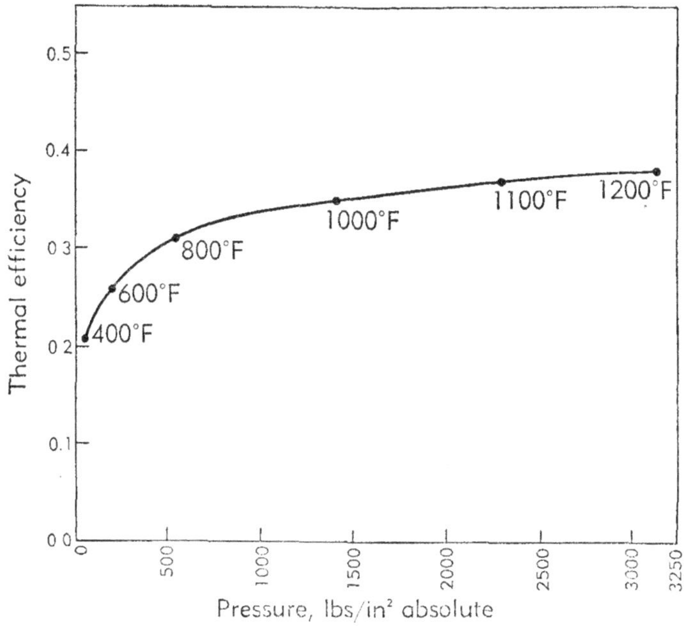

APPENDIX A

Variation of Efficiency with Temperature and Pressure. Special assumptions: All expansions are on single non-isentropic line of constant slope across Mollier steam diagram. Line passes through points (900°F temp., 875 lbs/in.²(abs.) pressure), and (29" Hg vacuum, 10% moisture), latter being exhaust conditions for all expansions.

These assumptions give results believed more practical than assumptions of perfect isentropic expansion or uniform enthalpy of exhaust.

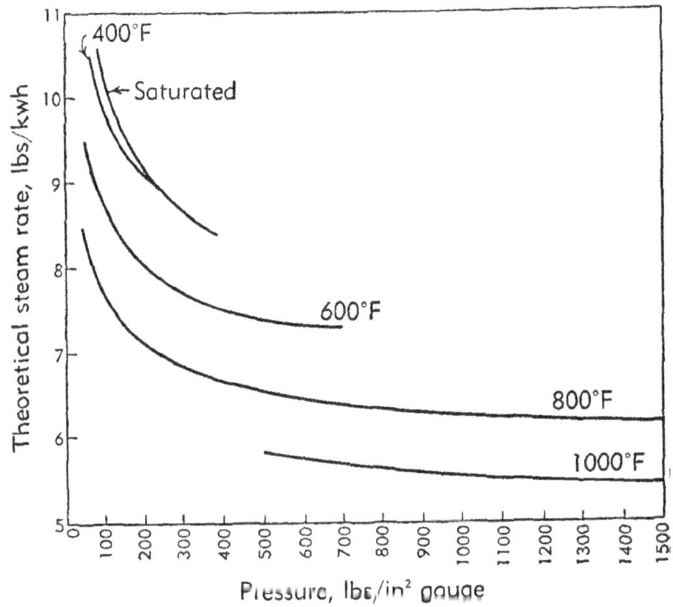

APPENDIX B

Variation of Theoretical Steam Rate with Temperature and Pressure. Curves were plotted directly from "Theoretical Steam Rate Tables" by J. H. Keenan and F. C. Keyes (A.S.M.E. 1938). They calculated the tables by the formula: Steam Rate = 3412.75 ÷ [(enthalpy, steam) — (enthalpy, exhaust)] and assumed perfect isentropic expansion. This assumption exaggerates advantages of pressure.

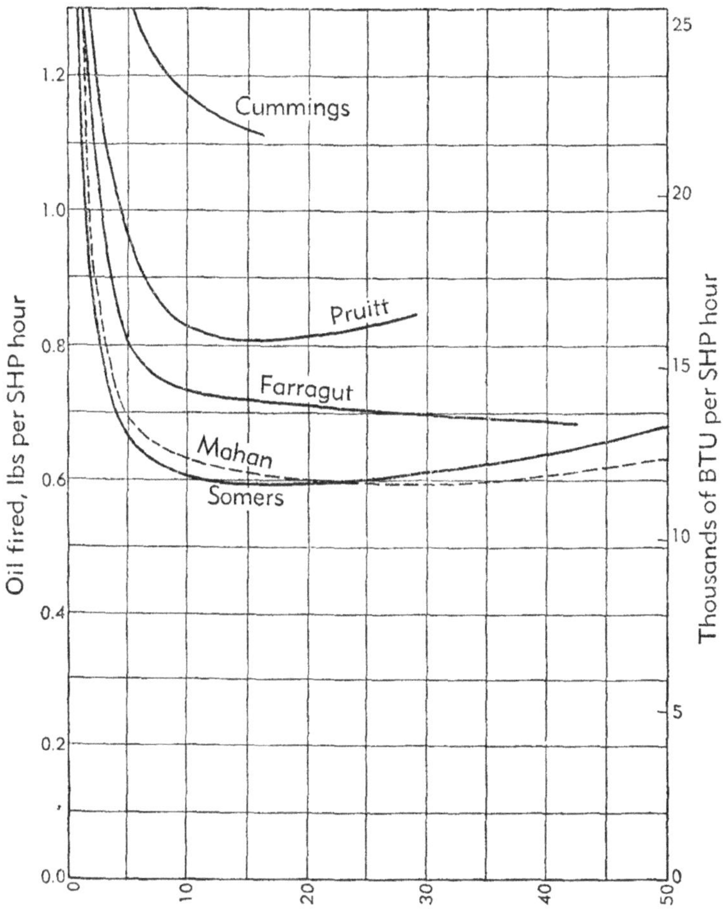

APPENDIX C

Comparative Economy of Destroyers. Data from official trials reduced to common basis of fuel having 19,500 BTU's per lb.

Ship	Year	Tons	Temp. °F.	Press., Gauge
Cummings	1913	1011	397	224
Pruitt	1920	1214	404	245
Farragut	1934	1594	571	430
Mahan	1936	1729	650	419
Somers	1937	2130	675	599

Temperature and pressure data are typical trial results at higher powers.

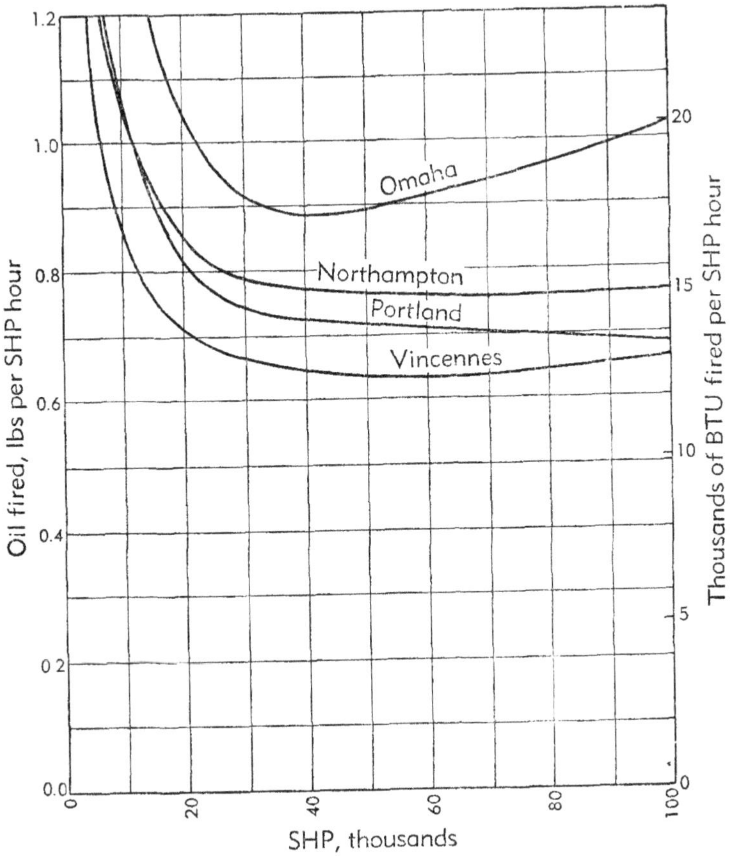

APPENDIX D

Comparative Economy of Cruisers. Data from official trials reduced to common basis of fuel having 19,500 BTU's per lb.

Ship	Year	Tons	Temp. °F.	Press., Gauge
Omaha	1923	7,286	411	265
Northampton	1930	10,653	422	302
Portland	1933	11,063	578	309
Vincennes	1936	11,558	664	449

Temperature and pressure data are typical trial results at higher powers.

APPENDIX E

BATTLESHIP ADVISORY BOARD

The following quotes from the report of the Battleship Advisory Board not only have a bearing on the subject matter of this autobiography but illustrate Naval thinking in 1938.

"Captain (Ormond L.) Cox[1] referred to his experience and responsibility in connection with new construction several years ago. He stated that at that time he had recommended to Admiral Harry E. Yarnell, Chief of the Bureau of Engineering, an initial steam condition of 250 pounds per square inch, saturated (steam), on account of saving 50 tons in machinery weights. As fuel did not count in [Limitation of Armament] Treaty weights, a saving in machinery weights permitted that much more armor or armament to be placed on a vessel. He expressed his personal opinion that reliability was greater for this steam condition than for higher pressures and temperatures; that first line vessels should be limited to 400 pounds per square inch and 200 degrees superheat and that 700 degrees Fahrenheit should be the limiting temperature aboard ship. . . . His opinion was that he would not advocate going back to former pressures and temperatures but would call a temporary halt to further increases.

"Rear Admiral S. M. Robinson added that the 400 pounds per square inch, 700 degrees Fahrenheit installation had been opposed by the same people that are now opposing the 600 pounds per square inch installation."

Rear Admiral Harry L. Brinser (Senior Member, Board of Inspection and Survey) stated that "too much stress had been placed on economy and not enough on reliability and ease and continuity of maintenance and operation . . . increased temperatures added to the complexity of the engine room; introduced greater fire hazards; led to difficulty with maintaining valves in operation condition . . . objected to these increases in pressure and temperature being applied to too many ships at once. He would prefer testing out two or three ships in service first. His general view was that it was desirable to sacrifice whatever might

[1] Author's Note: Captain Cox was chosen by the General Board later to conduct the General Board's hearings on high-pressure high-temperature steam.

be necessary for complete reliability. . . . He would be inclined to go back to the best types of Wartime [World War I] destroyers with 265 pounds per square inch, saturated steam, as these had proven themselves completely dependable in service.[2]

"Mr. Joseph Powell asked if he understood correctly that Admiral Brinser considered reliability more important than economy. The latter replied in the affirmative, in spite of constant demand from the forces afloat for increased steaming radius."

Admiral Brinser stated that the electrical load on destroyers was now greater than on the old battleships.

Rear Admiral William G. Dubose (Chief of the Bureau of Construction and Repair) preferred protection to speed.

It was stated that Vice Admiral Preece, Royal Navy (Engineer-in-Chief), preferred 350 pounds per square inch for destroyers, 400 pounds for other types, except battleships. His policy for battleships was not known.

"Admiral Arthur J. Hepburn [former Commander-in-Chief] stated that he would not open fire beyond 22,000 yards unless the enemy opened fire sooner. Vice Admiral Edward C. Kalbfus [later in charge of the War College] concurred in general but stated that 24,000 yards might probably prove to be the opening range. Vice Admiral Ernest J. King [later Commander-in-Chief and Chief of Naval Operations] believed that in general battle ranges would not exceed 18,000 to 20,000 yards.

"Admiral Joseph Strauss[3] [Senior Member of the Board] asked Admiral Hepburn if he attached much importance to attack by aircraft. Admiral Hepburn replied that he did not attach much importance to this form of attack as in his opinion the probability of hits was not high enough but agreed that the damage resulting from successful hits might be considerable. Admiral Strauss asked Vice Admiral King whether 10,000 feet was a practicable altitude from which to attack. Vice Admiral King replied that he believed the best altitude for bomb attacks

[2] Author's Note: These destroyers could not cross the Atlantic without refueling.

[3] Author's Note: At this time Admiral Strauss was suspicious of my motives and asked me if I were trying to be another Admiral Melville (a former Engineer-in-Chief and a figure of great power). After our high-pressure high-temperature ships were demonstrated to be successful, he took the occasion in public to congratulate me. Only one other gentleman among my opponents was gracious enough to do so—Mr. Arthur B. Homer, currently President of Bethlehem Steel.

was considerably higher, perhaps 15,000 feet. He stated that he did not have exact figures but thought that perhaps 5 per cent of hits would be obtained from this altitude. He stated that the improvement in bomb sights would in his opinion keep pace with the improvement in the fire of anti-aircraft guns. In reply to further questions by Admiral Strauss he stated that he considered the 500-pound bomb the best for all-around use at this time. He pointed out that our present patrol planes could each carry 12 bombs of this size, that each squadron contained 12 planes and that we now have 18 squadrons. These patrol planes carrying this load can attack from an altitude of 22,000 feet. He stated his own opinion that wartime hits would be about 2 per cent. He believed the best results were obtainable through team work with simultaneous attack by guns, torpedoes and bombs.

"Admiral Strauss inquired in regard to torpedo plane practice in the Fleet at present. Vice Admiral King stated that they now drop torpedoes from a height of 20 feet, but that with the new torpedoes now in production it would be possible to drop from a height of 100 feet. Admiral Hepburn stated that each plane could carry only one torpedo and that, as in the case of destroyers, torpedo planes would fire at the formation rather than at individual targets. He does not rate the torpedo plane very high. Vice Admiral King stated that in his opinion we should be prepared to use either bombs or torpedoes, that the new torpedo would have a range of 7,000 yards and that he believed its greatest field of usefulness was under conditions of low visibility."

Mr. William Francis Gibbs of Gibbs and Cox spoke in favor of high-pressure high-temperature steam and all the new American machinery and equipment.

Mr. John F. Metten (President, New York Shipbuilding Corporation) testified as follows:

"The adoption of 600 pounds boiler pressure for Naval work cannot be justified by any material gain in economy unless the steam temperature is raised to approximately 850 degrees Fahrenheit, and this temperature should not be used for Naval work of any character.

"The type of closed feed system now specified for the battleships is in my opinion undesirable. It has been installed in the similar destroyer installations referred to, where it has been one of the most obvious sources of crowding the engine room and complication of the feed system, as mentioned in the Naval

Trial Reports. Equally effective, lighter and simpler means of obtaining satisfactory deaeration of feed are available and should be employed.

"The type of divided furnace superheat control boilers specified for the (new) battleships is considered undesirable. They have been installed for the first time in destroyers that have just completed trials and must be regarded as experimental until they have had a year or two of continuous service operating with the Fleet. Laboratory tests under ideal conditions are not sufficient. The objection to them is the fact that an extremely undesirable boiler operation is introduced, which can and should be avoided, as the gains in operating economy are not sufficient to justify this disadvantage. The Standard Integral Superheater Boiler, which has been proven in service and with which all operating staffs of the Fleet are familiar, should be installed in capital ships until something unquestionably better is available.

"The proposed installation will result in questionable reliability, and in more congested machinery compartments and more complicated operating procedure than is desirable in a capital ship. Less congested compartments, simplified operation and considerable reduction in weight without sacrifice in economy or capacity can be obtained in a properly designed plant by reducing the pressure to 400 pounds per square inch and retaining a maximum temperature of 700 degrees eliminating the objectionable features previously referred to. A considerable reduction in cost would also result as well as a more reliable plant taking into consideration overall operation conditions as a combatant ship.

"The new British fleet is being built on the principle of each unit being added to the fleet having a known value. Proven reliability and simplicity of operation are given first place in their consideration of all Naval machinery installations. Any new developments are first tried out as experimental installations in unimportant vessels for several years under special service before risking wholesale adoption. I am of the opinion that this is a sound plan for us to follow in our new capital ships as well as other units of the Fleet. It is a plan that insures progress without undue risk. Battleships with unreliable machinery may be liabilities and not assets for National Defense."

APPENDIX F

ENGINEERING DATA

The improvement in engineering performance, after the high-pressure high-temperature steam fight, is strikingly demonstrated by a comparison of the figures of a pre-fight and a post-fight battleship. Displacement increased 43 per cent but horsepower increased almost 620 per cent. Total machinery weight increased 140 per cent but the ratio between horsepower and displacement increased about 470 per cent. Weight per horsepower decreased 67 per cent. Fuel expenditures decreased almost 37 per cent. While there was an increase in the complement of engineers of 45 per cent, each engineer produced five times as much horsepower. The light and power load rose over 500 per cent.

In the case of the destroyers, displacement increased almost 26 per cent, horsepower almost 17 per cent, total machinery weight 25 per cent. Fuel expenditures decreased 42 per cent.

APPENDIX G

ACHIEVEMENTS OF THE BUREAU OF ENGINEERING

There is a cliché that the next war always begins with the weapons and tactics of the last war. This, of course, is meant to be an indictment of the military for undesirable conservatism and lack of vision. I am not aware that the military are more tarred with these characteristics than any other element of society.

Let us examine what the Bureau of Engineering and the Naval Research Laboratory contributed to World War II that was not available in World War I.[1]

Surface Ships: Introduction of new American high-speed turbine design with forged rotors, welded casings, and double reduction gears; application of carbon molybdenum steel as required; controlled superheat boilers of greatly increased capacity with welded boiler drums and air casings; high-pressure high-temperature steam up to 1,200 pounds per square inch and 950°

[1] During this period the Naval Research Laboratory was under the Chief of the Bureau of Engineering.

Fahrenheit; deaerated feed-water systems; vastly improved methods for treatment of boiler water; alternating current for light and auxiliary power; emergency diesel-driven electric generators; flame-proof cable; safer switching apparatus for heavy currents; searchlights with stellite (metal) mirrors; acoustic and illumination surveys; elimination and reduction of noise on Naval vessels; standardization of illumination; dieselization of motor boats; search and fire control radar; homing devices for carrier- and land-based planes; landing devices; radio identification systems; IFF systems; air cooling for main generators of electric-drive ships (prototype for hydrogen-cooled generators of central power plants); improved gamma-ray photography and inclusion of gamma-ray, x-ray, and metallographic examinations in contracts; the first specifications in the country for sound movies; the first standard pedestals for main machinery so that equipment from any source could be installed easily aboard ship; sound-powered telephones which obviated the necessity for voice tubes; virtual elimination of intake noise of forced-draft blowers; introduction of turbine and motor-driven centrifugal pumps in place of steam reciprocating pumps; improved design of feed pumps to permit operation in parallel; development of instantaneous shaft revolution indicators; development of sectional fin-type fuel oil heaters; introduction of automatic dial telephone systems on ships; introduction of fluorescent lighting; introduction and development of the first forced-feed circulation boilers; large increase in capacity of fuel-oil burners for boilers; introduction of thyratron control; development of battle announcing systems; development of communication systems with gas masks; rotary-type underwater log; introduction on ships of welded steam pipes and VanStone joints, with elimination of steam pipe slip joints.

Submarines. Improved sonar, search, and fire control radar; choice of two American-designed multi-cylinder high-speed diesel engines; "pancake" vertical radial lightweight diesel (the prototype for the latest submarine radial diesel engine); electric-drive submarines; improvement in performance of storage batteries; development and introduction of magnetostriction sound device.

Miscellaneous. Radio-controlled airplanes, light armor, precipitation static control, hydrogen-peroxide fuel, potassium super oxide ("solid oxygen") for breathing, thermal diffusion method of concentrating U-235, anti-fouling paint, use of very high frequencies in communications, high-frequency direction finders, pulse altimeter and instruments to measure drift, anti-submarine detection devices (sono-radio buoy), introduction of Arthur D. Little Company's Kleinschmidt still for making potable fresh water from salt water with incredible efficiency, improvement in materials for withstanding salt-water corrosion.

Bureau of Engineering. Established a section for the estimating of costs and bids and for the scheduling and progress of ship building. In 1939 established a section for Research and Investigation, the first head of which was the late Captain Lybrand P. Smith. A planned program of research and experiment had already been established and a great deal of progress had already been made in the fields of packing (stuffing box) material, refractories, heat insulation, steel castings; full-speed full-load torque testing of each type of main reduction gear to improve tooth contact and reduce noise; replacing manually operated nozzle control valves on auxiliaries with automatic control; improving flexible couplings; designing auxiliary turbines to use superheated steam; improvement of switchboard and distribution systems; employment of a method devised by Dr. H. C. Hayes and Dr. Elias Klein of the Naval Research Laboratory to vibrate propeller blades in the investigation of stresses in thin propeller blades; creep tests of various metals at high temperatures; welded boiler tubes; improved design of main (steam) condensers; redesign of flanges, bolts, and gaskets; improvement in reducing and regulating valves; introduction of Freon refrigerating plants aboard ship; development of modern combined unit triple-effect distilling plants; establishment of a new piping and valve section in the Bureau to keep pace with rising temperatures and pressures and new alloys; pioneering with radiography among contractors; investigating the fabrication of turbine casings from rolled steel plate.

Navy Yards. Budgeting of Navy Yards began as early as 1927 at Puget Sound. The Bureau had gone far in successfully estimating costs of new vessels prior to the opening of bids. The Navy Yards were making much greater use of the Bureau's progress of work facilities and were submitting to the Bureau their own erection schedules on various types of vessels building which were coordinated into master schedules as required.

Because of the expanding building program and the increased maintenance due to the growing Fleet, every effort was made to increase the efficiency and sufficiency of the Yards in facilities, machine tool replacements, new and dieselized Yard craft, turret shops for Navy and private Yards, etc. This was particularly so in the case of the New York Navy Yard. The drydocks there could not dock our latest battleships. The Port of New York had no facilities to dock a ship the size of a *Queen Mary.* I convinced the leaders of the Yard workmen (union) and the civic leaders in Brooklyn that the future of their Yard depended on a drydock at the Bayonne, New Jersey, side of the Hudson, where a rock bottom was available. As a result of this and other construction, New York Navy Yard still retains its position as our largest and most important Navy Yard.

Radio Stations. The radio stations at Annapolis and Hawaii were considerably enlarged in area. New high-power low-frequency equipment was installed at Annapolis and the Canal Zone. Improvements were made in the communications establishments in the Mare Island, San Francisco, area. Preliminary investigations were made and a site selected at Barber's Point, T.H., for the establishment of a new communications center in the Hawaiian area. Extensive improvements were made in the antenna systems at Lualualei High-Power Low-Frequency station in Hawaii. All this to insure that we would have the best possible communications in the Pacific for the war we all knew was coming.

Sources of Supply. Every effort was made to increase sources of supply. I personally enlisted the interest of Foster Wheeler Company and Northern Battery Company. This gave us two sources for boilers and submarine batteries. At one time during the war, because Electric Storage Battery was otherwise importantly engaged, Northern Battery furnished about 60% of our submarine batteries. Sources of supply for machinery, radar, and equipment was enormously increased as noted earlier.

Quite a list of achievements for those EDO's "without pride of ancestry or hope of posterity."

This period represents the golden age of engineering in the Navy. There certainly was nothing remotely resembling it in the past, and I doubt that it will ever be duplicated in the future. There was in the Bureau of Engineering at this time literally a galaxy of engineers with brains, education, and a will to do. These loyal and able assistants gave everything they had to the amazing accomplishments of the final and last Bureau of Engineering:

Commander Roy W. Bruner, now Captain, retired
Commander Eliot H. Bryant, now Vice Admiral, retired
Commander Guy Chadwick, now Captain, retired
Commander David H. Clark, now Rear Admiral, retired
Commander Marshall M. Dana, now Captain
Commander Homer Davis, deceased
Commander Jennings B. Dow, now Commodore, retired
Commander Louis Dreller, now Rear Admiral, retired
Commander Peter W. Haas, Jr., now Rear Admiral
Commander James E. Hamilton, now Captain, retired
Captain James McR. Irish, now Rear Admiral, retired
Captain Claude A. Jones, later Rear Admiral, deceased
Commander Edgar P. Kranzfelder, now Rear Admiral, retired
Commander Paul F. Lee, now Rear Admiral, retired
Commander W. Durward Leggett, Jr., now Rear Admiral
Commander Earle W. Mills, now Vice Admiral, retired
Commander Roger W. Paine, now Rear Admiral, retired
Commander Wilbur J. Ruble, now Captain, retired
Commander Lisle F. Small, now Commodore, retired
Captain Harold T. Smith, now Rear Admiral, retired
Commander Lybrand P. Smith, later Captain, deceased
Commander Thorwald A. Solberg, now Rear Admiral, retired
Commander William E. Sullivan, later Rear Admiral, deceased

APPENDIX H

THE SECRETARY OF THE NAVY
WASHINGTON

13 February 1940

Dear Governor Scrugham:

I am very happy indeed to answer your letter relative to Admiral H. G. Bowen, USN, because I hold him in the highest regard, and may I say that I do *not* hold him responsible for the so-called top heavy destroyers you mentioned.

As Chief of the Bureau of Engineering, Admiral Bowen did an

outstanding and historically significant job. Many times, in the face of determined opposition, he effected far-reaching improvements in practically every phase of the work of the Bureau. I trust that his contributions to the Navy will be fittingly recognized when their importance becomes more generally appreciated. To adequately cover the scope of his work would require a lengthy paper.

Admiral Bowen's normal tour of duty terminated May 28, 1939. He was continued in an acting capacity as long as possible thereafter and then placed in charge of the Navy's scientific and technological work. In this duty he is still in a key position to make further vital contributions. I positively believe that he will do so. I confidently expect, too, that history will rate him as one of the most outstanding engineering officers of this generation.

Sincerely yours,

/s/ CHARLES EDISON

Honorable James G. Scrugham
House of Representatives

APPENDIX I

In the following press release the Bureau of Ships eulogizes the accomplishments of the Bureau of Engineering on whose ruins the Bureau of Ships was erected.

NAVY DEPARTMENT

HOLD FOR RELEASE
SUNDAY MORNING NEWSPAPERS
JUNE 13, 1943

NEW-DESIGN ENGINES GIVE "EDGE" TO AMERICAN FIGHTING SHIPS

Engineering developments made during peace time by the American power plant industry are being adapted to American fighting ships, giving them the "edge" over foreign vessels that many times means the difference between defeat and victory.

Along with the rapid improvement in gunnery and armor has come a corresponding advance in the means of propelling vessels at high rates of speed with the lowest possible fuel consumption.

The steam turbine that powers most ships is similar in principle to that which turns the huge generators that supply electric power throughout the country. On land the steam is generally produced with coal heat, while at sea oil generally is the fuel.

Power plant engineers for many years have been striving for greater and greater economy in the use of fuel and have developed machinery producing remarkable results—turbines operated with high-pressure and high-temperature steam.

These principles now are being developed on American warships where fuel economy is even more important than in land-located electric power stations. In less than a decade the fuel saving on large, modern warships has been increased by as much as 35 per cent in some cases compared with previous engine installations.

Not only does this result in valuable savings in scarce fuel oil and transportation facilities for such oil, but this extends the cruising radius of some vessels by a thousand miles or more and permits them to carry more armament. It also has the effect of creating more tonnage by keeping more ships on the line at all times.

Many years of study and experimental work in Naval and private laboratories preceded the application of modern electric power plant principles to American fighting ships. Only recently was the "go-ahead" signal given for the equipping of American Naval vessels with this modern machinery, one of the greatest changes that ever took place in marine engineering. The work proceeded under the direction of the Bureau of Engineering.

The steam turbine is one of the most simple of engines. It operates on the same principle as the windmill, where a breeze striking the blades causes the shaft to rotate. In the steam turbine jets of steam are directed against the turbine blades. The arrangement, the number, and the strength of the turbine assembly, together with the direction and force of the steam jet, largely determine the speed of the turbine, its power, and its efficiency.

The most modern ship installations of this type superheat the steam to about 925 degrees Fahrenheit, at pressures of nearly a quarter ton per square inch, before it enters the turbine. At this temperature the steam is so hot it heats the pipes to a dull red and necessitates the use of highly specialized equipment.

New types of steel and other metals were developed to handle the high pressures and temperatures.

The degree of superheat used is controlled by boiler dampers in the central heating power plants of the United States. This feature is very useful in boosting the efficiency of the plant, especially when operating at lower power. On account of the limited space aboard ships, however, it was impractical to use boiler dampers. Consequently the Bureau of Engineering developed superheat control boilers, the first of their kind, which can control superheat so accurately that changes cannot be recorded satisfactorily by a mercury thermometer because of the inertia of the mercury. The changes are so rapid that they can only be followed by a pyrometer.

The speed of the turbine, however, is so great that it must be reduced by 10 to 12 times through newly developed reducing gears before the power is applied to the propeller shaft. The double reduction gears now used are very superior to the single reduction gears which they replaced, and the tooth pressures and bearing pressures are lower. The steam is used several times before it has spent its force by introducing it through several turbines operated in series. The ship's forward or reverse progress is controlled merely by increasing or decreasing the steam supply and the number of turbines into which the steam is introduced.

Another improvement over older types of equipment is a new means of excluding air from the inside of steam systems. As the temperature of feed water is increased, its affinity for oxygen is stepped up. Oxygen is a great destroyer of boiler tubes and drums and with the introduction of high-temperature high-pressure steam came the need of a much more satisfactory deaeration system than had previously existed on Naval vessels. The Bureau of Engineering, realizing that the oxygen in feed systems was a more serious affair aboard ships than in central power plants, successfully evolved what is probably one of the most efficient systems for removing oxygen from feed water, using much of the experience gained in the steam power plants in the United States.

When the expansion of the Navy started in 1933, the firm of Gibbs and Cox, which had considerable experience designing high-pressure high-temperature equipment on merchant ships, was brought into the field by the Bath Iron Works, United Shipbuilding and Drydock Company, and the Federal Shipbuilding and Drydock Company. In the *Mahan* class of destroyers, then building, it was attempted to embody a design utilizing steam at 400 pounds pressure and 850 degrees Fahrenheit, which later was cut back to 700 degrees Fahrenheit.

The design included integral superheat boilers with economizers but not air casings. It was distinguished also from earlier designs in that the cruising or high-pressure turbine was directly coupled and always engaged. The main gears were the new double reduction type and a deaerated feed water system was employed. This design represented the first time that efficient high-speed turbines with double reduction gears and direct-coupled cruising turbines had ever been installed in a Naval ship. These ships represented a complete American design in all respects.

The *Somers*, contracted for in 1934, designed by Gibbs and Cox, and built in Federal Shipbuilding and Drydock Company, had all of these innovations together with a steam condition of 600 pounds and 700 degrees Fahrenheit, as well as new superheat control boilers which also incorporated a double casing within which the air necessary to support combustion was confined, having a definite military asset in case of gas attack. As finally built, the *Somers* was operated on a steam pressure of 600 pounds and 850 degrees Fahrenheit, which is 369 degrees of superheat.

As a result of the success of the *Somers*, the Bureau of Engineering raised the steam pressure on the battleships *North Carolina* and *Washington*, building at the New York and Philadelphia Navy Yards, from 700 degrees Fahrenheit to 850 degrees Fahrenheit. They had already been designed for 600 pounds pressure.

Since the last war, the size, scope, and influence of the American industrial research laboratories has increased by leaps and bounds. Gear cutting has become a highly specialized endeavor, and the turbine companies, particularly the General Electric Company and Westinghouse Electric and Manufacturing Company, have poured millions of dollars into the development of steam turbines. Much attention to engine research also was given by Babcock and Wilcox, Foster Wheeler Corporation, and the DeLaval Company.

All of the improvements incident to the introduction of high-pressure high-temperature have worked out most satisfactorily. The machinery has stood up well under the most severe conditions imaginable, namely—convoy duty off Iceland and combat action in the tropics. It has shown itself particularly well adaptable to operation by green crews. It became standard for surface combatant vessels beginning with 1938.

In 1939, the four-stack destroyer, *Dahlgren*, was completely re-engined and re-boilered. This ship now uses steam at 1300 pounds pressure and 925 degrees Fahrenheit in the main engines, which is reported to result in a still further improvement of ten per cent in economy. While nothing has been done with this design in the last four years, it is understood that the Bureau of Ships will follow up this design in the near future, as the material situation permits.

APPENDIX J

DIFFICULTIES IN INTRODUCING ELECTRIC DRIVE IN THE NAVY

About 1909 William Le Roy Emmet, Chief Engineer of the General Electric Company, conceived the idea of introducing turbo-electric drive aboard ship. Emmet was a graduate of the U.S. Naval Academy who left the Navy after completing his two-years' cruise as a passed Midshipman. In 1910 Charles A. Parsons created somewhat of a sensation when he introduced the reduction gear aboard the steamship *Vespasian*.[1] Both men knew that the propeller was most efficient at relatively low speeds and the turbine at high speeds. Both men were looking for a device to reduce the high speed of the turbine to the low speed of the propeller. Emmet sought to do this by using high-speed turbines to drive electric generators and use the electricity generated to run motors attached to the propeller shaft. He attempted to push electric drive in the Navy and his report on these activities has a familiar ring and reminds one so much of the high-pressure high-temperature controversy. I quote as follows:[2]

[1] See Chapter II—High-Pressure High-Temperature, page 49.

[2] *The Autobiography of An Engineer*, by William Le Roy Emmet, published by The American Society of Mechanical Engineers, New York, 1940, pp. 158-160.

"My opinions [regarding electric drive], however, carried little or no weight in the Navy, and the shipbuilding fraternity, whose relations with the Navy were very intimate, were hotly opposed to such an innovation, the merits of which they did not understand and which if adopted would take the manufacture of propelling machinery out of their yards. Many of the leading men in the shipyards had been naval constructors and all the naval constructors had for years been selected from the cadets at the Naval Academy who had stood at or near the top of their classes.

"From what I have said in previous chapters it may be judged that I do not consider that the scholastic standard alone affords a sufficient basis for the selection of men, particularly in the matter of engineering, which calls for particular qualities which the average good scholar may not have. In the Navy the standard of scholarship is that which governs class standing and promotion and is consequently of much importance and attracts much attention. The good scholars tend to be started at the top of everything and I think that that is one reason why they are sometimes not so good in after life.

"I say all this to show a difficulty which I faced when I sought to force a radical engineering change upon the Navy. I had been a particularly undistinguished student at the Naval Academy and had many years before been mustered out of the Navy on account of my deficiency in class standing. I had some friends and classmates in the Navy and a few of them were officers on the General Board, but among these I was in the situation of a prophet in his own country and I fear I was looked upon rather as a joke than as a serious factor in marine engineering. One distinguished captain who had stood at the head of his class at Annapolis and who as an instructor there had experienced some unimpressive relations with me said in the General Board, as I was afterwards told, that he thought it quite impossible that a man of my type could produce anything valuable in such a connection.

"For such reasons as I have suggested and for those common in the introduction of novelties my task in the promotion of electric drive was not an easy one. I talked and wrote and lectured for about seven years before the full result was accomplished in the Navy. . . ."

Westinghouse entered the competition between electric reduction and gear reduction with the Melville-McAlpine reduction gear.[3] Admiral George W. Melville was formerly Chief of the Bureau of Steam Engineering and John H. McAlpine was with the Westinghouse Corporation.

Soon afterwards the collier *Neptune* was equipped with this gear. This gave Emmet of General Electric the leverage to obtain a contract to equip a sister collier, *Jupiter*, with electric (reduction) drive. The *Jupiter* later as the *Langley* became famous as our first aircraft carrier. Another sister ship, *Cyclops*, was equipped with reciprocating engines as had been planned originally for all three.

As a matter of record, "The *Cyclops* and *Neptune* were found to be materially inferior in fuel economy and the *Neptune* showed a greatly inferior record in the matter of availability. The *Cyclops* was lost mysteriously while bringing a cargo of ore from South America."[4]

The next electric drive ship was the battleship *New Mexico* equipped with Emmet's General Electric design. About this time plans were being prepared for some very large battle cruisers of which only two were built and these, the *Saratoga* and *Lexington*, finally emerged as aircraft carriers. These ships were designed for 34 knots and actually developed over 210,000 horsepower.

"The Navy Department decided to use electric drive in these ships and a very active propaganda was begun by shipbuilders and others to have the plan given up. Several senators were induced to interest themselves in this effort. An attempt was made to organize a lot of prominent engineers in protest and several daily papers, both in New York and Chicago, printed whole pages of argument, denunciation, and profuse illustration. My only response to this campaign was to write an article for the *Electrical World* which was intended to back up the decision of the naval authorities. The article was a good one and presumably tended to discourage our opponents but support was not much needed by the Navy because Admiral Robert S. Griffin, the Chief of the Bureau of Steam Engineering, had good reasons for his decision and was not the kind of man to change on account of such pressure as was brought to bear upon him."[5]

[3] See Chapter II, p. 32. [4] Emmet, *op.cit.*, p. 164. [5] *Ibid.*, pp. 170-171.

The job of designing and building the machinery for these ships was a record for the electrical industry. Each of the eight motors in each ship was five or six times as large as the largest motors previously built. In the 1920's Lake Cushman, which supplied water to the municipal hydroelectric plant of the city of Tacoma, Washington, went practically dry. Upon request, the *Lexington* tied up to a wharf at Tacoma and supplied the entire city with light and power. It is interesting to note that Stone and Webster once investigated this same site for Puget Sound Power and Light and rejected it because of unsatisfactory annual rainfall figures.

"The work on this large warship machinery was stopped during the period of the war and after the Conference for Limitation of Armaments two of the four cruiser equipments which we had on order were cancelled and two ordered completed for the *Saratoga* and *Lexington,* so that the first of these ships was not run until about nine years after we took the contract. We were paid something for cancellation and storage but the machinery crowded our shops for years and, being finally paid for practically at prewar prices, this great enterprise was not a very profitable one for the General Electric Company. My experience is that the United States government is a very difficult customer to deal with. Everyone means well but no one will take the responsibility of making any kind of concession, no matter what the circumstances."[6]

During World War I reduction gears were extensively used for destroyers, General Electric alone contributing machinery of 28,000 horsepower each for about 50 destroyers.

Routine workers are a necessary part of life. They are the people for whom organization charts, time clocks, job classification sheets and such are devised, but "enterprises like these ship undertakings, which involve the promotion of new things and methods, must always be to a great extent one-man jobs, and my definiteness and fixity of purpose was the important factor, but I had much able assistance and cooperation in the ship activity as in others."[7]

The earliest converts to electric drive in the Navy were Captain S. S. Robison, later Commander in Chief of the Battle Fleet and the U.S. Fleet, and Lieutenant S. M. Robinson, later Chief of the Bureau of Ships.

[6] *Ibid.*, pp. 171-172. [7] *Ibid.*, pp. 172-173.

This experience with electric drive seems to have taught the Navy Department and the General Board[8] absolutely nothing. My antagonists in high-pressure high-temperature steam were exactly the same as Emmet's in electric drive. It was perfectly ridiculous to think that 210,000 horsepower in four shafts could have been handled by the totally inadequate reduction gears of that period. Years after, and after gear hobbing became a highly developed specialty on a weight and space basis, electric drive was superseded by the double reduction gear.

[8] The General Board was abolished 10 March 1951. De mortuis nihil nisi bonum.

Index

INDEX

GPSR Authorized Representative: Easy Access System Europe - Mustamäe tee 50, 10621 Tallinn, Estonia, gpsr.requests@easproject.com

www.ingramcontent.com/pod-product-compliance
Lightning Source LLC
LaVergne TN
LVHW052222200926
848481LV00001B/179